Formulas and Calculations for Drilling Operations Second Edition

Scrivener Publishing
100 Cummings Center, Suite 541J
Beverly, MA 01915-6106

Publishers at Scrivener
Martin Scrivener (martin@scrivenerpublishing.com)
Phillip Carmical (pcarmical@scrivenerpublishing.com)

Formulas and Calculations for Drilling Operations
Second Edition

James G. Speight

Scrivener Publishing

WILEY

This edition first published 2018 by John Wiley & Sons, Inc., 111 River Street, Hoboken, NJ 07030, USA and Scrivener Publishing LLC, 100 Cummings Center, Suite 541J, Beverly, MA 01915, USA

For more information about Scrivener publications please visit www.scrivenerpublishing.com.

Wiley Global Headquarters
111 River Street, Hoboken, NJ 07030, USA

For details of our global editorial offices, customer services, and more information about Wiley products visit us at www.wiley.com.

Library of Congress Cataloging-in-Publication Data
ISBN 978-1-119-08362-7

Cover design by Kris Hackerott

Set in size of 11pt and Minion Pro by Exeter Premedia Services Private Ltd., Chennai, India

10 9 8 7 6 5 4 3 2 1

Contents

Preface

Drilling engineers design and implement procedures to drill wells as safely and economically as possible. Drilling engineers are often degreed as petroleum engineers, although they may come from other technical disciplines (such as mechanical engineering, geology, or chemical engineering) and subsequently be trained by an oil and gas company. The drilling engineering also may have practical experience as a rig hand or mud-logger or mud engineer.

The drilling engineer, whatever his/her educational background, must work closely with the drilling contractor, service contractors, and compliance personnel, as well as with geologists, chemists, and other technical specialists. The drilling engineer has the responsibility for ensuring that costs are minimized while getting information to evaluate the formations penetrated, protecting the health and safety of workers and other personnel, and protecting the environment. Furthermore, to accomplish the task associated with well drilling and crude oil (or natural gas production) it is essential that the drilling engineers has a convenient source of references to definitions, formulas and examples of calculations.

This Second Edition continues as an introductory test for drilling engineers, students, lecturers, teachers, software programmers, testers, and researchers. The intent is to provide basic equations and formulas with the calculations for downhole drilling. In addition, where helpful, example calculations are included to show how the formula can be employed to provide meaningful data for the drilling engineer.

The book will provide a guide to exploring and explaining the various aspects of drilling engineering and will continue to serve as a tutorial guide for students, lecturers, and teachers as a solution manual and is a source for solving problems for drilling engineers.

For those users who require more details of the various terms and/or explanation of the terminology, the book also contain a comprehensive bibliography and a Glossary for those readers/users who require an explanation of the various terms. There is also an Appendix that contains valuable data in a variety of tabular forms that the user will find useful when converting the various units used by the drilling engineer.

Dr. James Speight,
Laramie, Wyoming.
January 2018.

1

Standard Formulas and Calculations

1.01 Abrasion Index

The abrasion index (sometimes referred to as the *wear index*) is a measure of equipment (such as drill bit) wear and deterioration. At first approximation, the wear is proportional to the rate of fuel flow in the third power and the maximum intensity of wear in millimeters) can be expressed:

$$\delta pl = \alpha\ \eta\ k\ m\ \omega^3\ \tau$$

δpl – maximum intensity of plate wear, mm.
α – abrasion index, mm s^3/g h.
η – coefficient, determining the number of probable attacks on the plate surface.
k – concentration of fuel in flow, g/m^3.
m – coefficient of wear resistance of metal;
w – velocity of fuel flow, meters/sec.
τ – operation time, hours.

The resistance of materials and structures to abrasion can be measured by a variety of test methods (Table 1.1) which often use a specified abrasive or other controlled means of abrasion. Under the conditions of the test, the results can be reported or can be compared items subjected to similar tests. Theses standardized measurements can be employed to produce two sets of data: (1) the *abrasion rate*, which is the amount of mass lost per

Table 1.1 Examples of Selected ASTM Standard Test Method for Determining Abrasion*.

ASTM B611	Test Method for Abrasive Wear Resistance of Cemented Carbides
ASTM C131	Standard Test Method for Resistance to Degradation of Small-Size Coarse Aggregate by Abrasion and Impact in the Los Angeles Machine
ASTM C535	Standard Test Method for Resistance to Degradation of Large-Size Coarse Aggregate by Abrasion and Impact in the Los Angeles Machine
ASTM C944	Standard Test Method for Abrasion Resistance of Concrete or Mortar Surfaces by the Rotating-Cutter Method
ASTM C1353	Standard Test Method for Abrasion Resistance of Dimension Stone Subjected to Foot Traffic Using a Rotary Platform, Double-Head Abraser
ASTM D 2228	Standard Test Method for Rubber Property – Relative Abrasion Resistance by the Pico Abrader Method
ASTM D4158	Standard Guide for Abrasion Resistance of Textile Fabrics, see Martindale method
ASTM D7428	Standard Test Method for Resistance of Fine Aggregate to Degradation by Abrasion in the Micro-Deval Apparatus
ASTM G81	Standard Test Method for Jaw Crusher Gouging Abrasion Test
ASTM G105	Standard Test Method for Conducting Wet Sand/Rubber Wheel Abrasion Tests
ASTM G132	Standard Test Method for Pin Abrasion Testing
ASTM G171	Standard Test Method for Scratch Hardness of Materials Using a Diamond Stylus
ASTM G174	Standard Test Method for Measuring Abrasion Resistance of Materials by Abrasive Loop Contact

*ASTM International, West Conshohocken, Pennsylvania; test methods are also available from other standards organizations.

1000 cycles of abrasion, and (2) the *normalized abrasion rate*, which is also called the *abrasion resistance index* and which is the ratio of the abrasion rate (i.e., mass lost per 1000 cycles of abrasion) with the known abrasion rate for some specific reference material.

1.02 Acid Number

The *acid number* (*acid value, neutralization number, acidity*) is the mass of potassium hydroxide (KOH) in milligrams that is required to neutralize one gram of the substance (ASTM D664, ASTM D974).

$$AN = (V_{eq} - b_{eq})N(56.1/W_{oil})$$

V_{eq} is the amount of titrant (ml) consumed by the crude oil sample and 1 ml spiking solution at the equivalent point, b_{eq} is the amount of titrant (ml) consumed by 1 ml spiking solution at the equivalent point, and 56.1 is the molecular weight of potassium hydroxide.

1.03 Acidity and Alkalinity

pH is given as the negative logarithm of $[H^+]$ or $[OH^-]$ and is a measurement of the acidity of a solution and can be compared by using the following:

$$pH = -\log([H^+]$$

$$pH = -\log([OH^-]$$

$[H^+]$ or $[OH^-]$ are hydrogen and hydroxide ion concentrations, respectively, in moles/litter. Also, at room temperature, pH + pOH = 14. For other temperatures:

$$pH + pOH = pK_w$$

K_w is the ion product constant at that particular temperature. At room temperature, the ion product constant for water is 1.0×10^{-14} moles/litter (mol/L or M). A solution in which $[H^+=] > [OH^-]$ is acidic, and a solution in which $[H^+=] < [OH^-]$ is basic (Table 1.2).

Table 1.2 Ranges of Acidity and Alkalinity.

pH	*[H+]*	*Property*
<7	$>1.0 \times 10^{-7}$ M	Acid
7	1.0×10^{-7} M	Neutral
>7	$<1.0 \times 10^{-7}$ M	Basic

1.04 Annular Velocity

Three main factors affecting annular velocity are size of hole (bigger ID), size of drill pipe (smaller OD) and pump rate. Thus:

$$\text{Annular velocity, ft/min} = \text{Flow rate, bbl/min} \div \text{annular capacity, bbl/ft}$$

For example, with a flow rate of 10 bbl/min and an annular capacity of 0.13 bbl/ft, the annular velocity is:

$$10 \text{ bbl/min} \div 0.13 \text{ bbl/ft which is } 76.92 \text{ ft/min.}$$

Other formulas include:

$$\text{Annular velocity, ft/min} = (24.5 \times Q) \div (Dh^2 - Dp^2)$$

where Q is the flow rate in gpm, Dh is inside diameter of casing or hole size in inches, and Dp is outside diameter of pipe, tubing or collars in inch. Thus, for a flow rate of 800 gpm, a hole size of 10 inches, a drill pipe OD of 5 inches, the annular velocity is

$$\text{Annular velocity} = (24.5 \times 800) \div (10^2 - 5^2) = 261 \text{ ft/min}$$

Another formula used is:

$$\text{Annular Velocity, ft/min} = \text{Flow rate (Q), bbl/min} \times 1029.4 \div (Dh^2 - Dp^2).$$

Thus, for a flow rate equal to 13 bbl/min, a hole size of 10 inches, and a drill pipe OD of 5 inches, the annular velocity is:

$$13 \text{ bbl/min} \times 1029.4 \div (102 - 52) = 178.43 \text{ ft/min}$$

1.05 Antoine Equation

The Antoine equation is a correlation used for describing the relation between vapor pressure and temperature for pure components. The Antoine constants A, B, and C (Table 1.3) are component specific constants for the Antoine equation:

$$\log_{10}P = A - (B/C + T)$$

$$T = [B/(A - \log_{10}P) - C$$

P is the vapor pressure, mm Hg, and *T* is the temperature, °C.

1.06 API Gravity – Kilograms per Liter/Pounds per Gallon

The American Petroleum Institute gravity (API gravity) is a measure of how heavy or light a petroleum liquid is compared to water: if the API gravity is greater than 10, it is lighter than water and floats on water. On the other hand, if the API gravity is less than 10, it is heavier than water and sinks. The formula to calculate API gravity from the specific gravity is:

$$\text{API gravity} = (141.5/\text{specific gravity}) - 131.5$$

Conversely, the specific gravity of petroleum liquids can be derived from their API gravity value by the equation:

$$\text{Specific gravity at } 60\ °\text{F} = 141.5/(\text{API gravity} + 31.5)$$

Table 1.3 Example of the Antoine Constants.

	A	B	C	T_{min}, °C	T_{max}, °C
Water	8.07131	1730.63	233.426	1	100
Water	8.14019	1810.94	244.485	99	374
Ethanol	8.20417	1642.89	230.3	−57	80
Ethanol	7.68117	1332.04	199.2	77	243

Using the API gravity, it is possible to calculate the approximate number of of crude oil per metric ton. Thus:

$$\text{Barrels of crude oil per metric ton} = (\text{API gravity} + 131.5)/(141.5 \times 0.159)$$

The relationship between the API gravity of crude oil and kilograms per liter or pounds per gallon is presented in the table (Table 1.4) below.

Table 1.4 API Gravity Conversion to Kilograms per Liter/Pounds per Gallon

API gravity	Specific gravity	Kilograms per liter	Pounds per gallon
1	1.0679	1.0658	8.8964
1.5	1.0639	1.0618	8.863
2	1.0599	1.0578	8.8298
2.5	1.056	1.0539	8.7968
3	1.052	1.0499	8.7641
3.5	1.0481	1.0461	8.7317
4	1.0443	1.0422	8.6994
4.5	1.0404	1.0384	8.6674
5	1.0366	1.0346	8.6357
5.5	1.0328	1.0308	8.6042
6	1.0291	1.027	8.5729
6.5	1.0254	1.0233	8.5418
7	1.0217	1.0196	8.511
7.5	1.018	1.0159	8.4804
8	1.0143	1.0123	8.45
8.5	1.0107	1.0087	8.4198
9	1.0071	1.0051	8.3898
9.5	1.0035	1.0015	8.3601
10	1	0.998	8.3306
10.5	0.9965	0.9945	8.3012
11	0.993	0.991	8.2721
11.5	0.9895	0.9875	8.2432

(Continued)

Table 1.4 Cont.

API gravity	Specific gravity	Kilograms per liter	Pounds per gallon
12	0.9861	0.9841	8.2144
12.5	0.9826	0.9807	8.1859
13	0.9792	0.9773	8.1576
13.5	0.9759	0.9739	8.1295
14	0.9725	0.9706	8.1015
14.5	0.9692	0.9672	8.0738
15	0.9659	0.9639	8.0462
15.5	0.9626	0.9607	8.0189
16	0.9593	0.9574	7.9917
16.5	0.9561	0.9542	7.9647
17	0.9529	0.951	7.9379
17.5	0.9497	0.9478	7.9112
18	0.9465	0.9446	7.8848
18.5	0.9433	0.9414	7.8585
19	0.9402	0.9383	7.8324
19.5	0.9371	0.9352	7.8064
20	0.934	0.9321	7.7807
20.5	0.9309	0.9291	7.7551
21	0.9279	0.926	7.7297
21.5	0.9248	0.923	7.7044
22	0.9218	0.92	7.6793
22.5	0.9188	0.917	7.6544
23	0.9159	0.914	7.6296
23.5	0.9129	0.9111	7.605
24	0.91	0.9081	7.5805
24.5	0.9071	0.9052	7.5562
25	0.9042	0.9023	7.5321
25.5	0.9013	0.8995	7.5081
26	0.8984	0.8966	7.4843
26.5	0.8956	0.8938	7.4606

(Continued)

Table 1.4 Cont.

API gravity	Specific gravity	Kilograms per liter	Pounds per gallon
27	0.8927	0.891	7.4371
27.5	0.8899	0.8882	7.4137
28	0.8871	0.8854	7.3904
28.5	0.8844	0.8826	7.3673
29	0.8816	0.8799	7.3444
29.5	0.8789	0.8771	7.3216
30	0.8762	0.8744	7.2989
30.5	0.8735	0.8717	7.2764
31	0.8708	0.869	7.254
31.5	0.8681	0.8664	7.2317
32	0.8654	0.8637	7.2096
32.5	0.8628	0.8611	7.1876
33	0.8602	0.8585	7.1658
33.5	0.8576	0.8559	7.1441
34	0.855	0.8533	7.1225
34.5	0.8524	0.8507	7.101
35	0.8498	0.8482	7.0797
35.5	0.8473	0.8456	7.0585
36	0.8448	0.8431	7.0375
36.5	0.8423	0.8406	7.0165
37	0.8398	0.8381	6.9957
37.5	0.8373	0.8356	6.975
38	0.8348	0.8331	6.9544
38.5	0.8324	0.8307	6.934
39	0.8299	0.8283	6.9136
39.5	0.8275	0.8258	6.8934
40	0.8251	0.8234	6.8733
40.5	0.8227	0.821	6.8533
41	0.8203	0.8186	6.8335
41.5	0.8179	0.8163	6.8137

(Continued)

Table 1.4 Cont.

API gravity	Specific gravity	Kilograms per liter	Pounds per gallon
42	0.8156	0.8139	6.7941
42.5	0.8132	0.8116	6.7746
43	0.8109	0.8093	6.7551
43.5	0.8086	0.807	6.7358
44	0.8063	0.8047	6.7167
44.5	0.804	0.8024	6.6976
45	0.8017	0.8001	6.6786
45.5	0.7994	0.7978	6.6597
46	0.7972	0.7956	6.641
46.5	0.7949	0.7934	6.6223
47	0.7927	0.7911	6.6038
47.5	0.7905	0.7889	6.5853
48	0.7883	0.7867	6.567
48.5	0.7861	0.7845	6.5487
49	0.7839	0.7824	6.5306
49.5	0.7818	0.7802	6.5126
50	0.7796	0.7781	6.4946
50.5	0.7775	0.7759	6.4768
51	0.7753	0.7738	6.459
51.5	0.7732	0.7717	6.4414
52	0.7711	0.7696	6.4238
52.5	0.769	0.7675	6.4064
53	0.7669	0.7654	6.389
53.5	0.7649	0.7633	6.3717
54	0.7628	0.7613	6.3546
54.5	0.7608	0.7592	6.3375
55	0.7587	0.7572	6.3205
55.5	0.7567	0.7552	6.3036
56	0.7547	0.7532	6.2868
56.5	0.7527	0.7512	6.2701

(Continued)

Table 1.4 Cont.

API gravity	Specific gravity	Kilograms per liter	Pounds per gallon
57	0.7507	0.7492	6.2534
57.5	0.7487	0.7472	6.2369
58	0.7467	0.7452	6.2204
58.5	0.7447	0.7432	6.2041
59	0.7428	0.7413	6.1878
59.5	0.7408	0.7394	6.1716
60	0.7389	0.7374	6.1555
60.5	0.737	0.7355	6.1394
61	0.7351	0.7336	6.1235
61.5	0.7332	0.7317	6.1076
62	0.7313	0.7298	6.0918
62.5	0.7294	0.7279	6.0761
63	0.7275	0.7261	6.0605
63.5	0.7256	0.7242	6.045
64	0.7238	0.7223	6.0295
64.5	0.7219	0.7205	6.0141
65	0.7201	0.7187	5.9988
65.5	0.7183	0.7168	5.9836
66	0.7165	0.715	5.9685
66.5	0.7146	0.7132	5.9534
67	0.7128	0.7114	5.9384
67.5	0.7111	0.7096	5.9235
68	0.7093	0.7079	5.9086
68.5	0.7075	0.7061	5.8939
69	0.7057	0.7043	5.8792
69.5	0.704	0.7026	5.8645
70	0.7022	0.7008	5.85
70.5	0.7005	0.6991	5.8355
71	0.6988	0.6974	5.8211
71.5	0.697	0.6957	5.8068

(Continued)

Table 1.4 Cont.

API gravity	Specific gravity	Kilograms per liter	Pounds per gallon
72	0.6953	0.6939	5.7925
72.5	0.6936	0.6922	5.7783
73	0.6919	0.6905	5.7642
73.5	0.6902	0.6889	5.7501
74	0.6886	0.6872	5.7361
74.5	0.6869	0.6855	5.7222
75	0.6852	0.6839	5.7083

Table 1.5 API Gravity and Sulfur Content of Selected Heavy Oils.

	API	Sulfur % w/w
Bachaquero	13.0	2.6
Boscan	10.1	5.5
Cold Lake	13.2	4.1
Huntington Beach	19.4	2.0
Kern River	13.3	1.1
Lagunillas	17.0	2.2
Lloydminster	16.0	2.6
Lost Hills	18.4	1.0
Merey	18.0	2.3
Midway Sunset	12.6	1.6
Monterey	12.2	2.3
Morichal	11.7	2.7
Mount Poso	16.0	0.7
Pilon	13.8	1.9
San Ardo	12.2	2.3
Tremblador	19.0	0.8
Tia Juana	12.1	2.7
Wilmington	17.1	1.7
Zuata Sweet	15.7	2.7

Table 1.6 API Gravity at Observed Temperature Versus API Gravity at 60 °F

Observed temperature (°F)	18.0	19.0	20.0	21.0	22.0	23.0	24.0	25.0	26.0	27.0
70	17.5	I8.4	19.4	20.4	21.4	22.4	23.4	24.4	25.4	26.3
75	17.2	I8.2	19.1	20.1	21.1	22.1	23.1	24.1	25.0	26.0
80	16.9	17.9	ia.9	19.a	20.a	21.a	22.S	23.7	24.7	25.7
85	16.6	17.6	ia.6	19.6	20.5	21.5	22.5	23.4	24.4	25.4
90	16.4	17.3	is.3	19.3	20.2	21.2	22.2	23.1	24.1	25.1
95	16.1	17.1	13.0	19.0	20.0	20.9	21.9	22.S	23.S	24.S
100	15.9	16.a	17.8	18.7	19.7	20.6	21.6	22.5	23.5	24.4
105	15.6	16.5	17.5	18.7	19.4	20.3	21.3	22.2	23.2	24.1
110	15.3	16.3	17.2	18.2	1 9.1	20.1	21.0	21.9	22.9	23.B
115	15.1	16.0	17.0	17.9	la.a	19.a	20.7	21.6	22.6	23.5
120	14.8	15.8a	16.7	17.6	ia.6	19.5	20.4	21.3	22.3	23.2
125	14.6	15.5	16.4	17.4	ia.3	19.2	20.1	21.1	22.0	22.9
130	14.3	15.2	16.2	17.4	la.o	IS.9	19.9	20.S	21.7	22.6
135	14.1	15.0	15.9	16.B	17.7	IS.7	19.6	20.5	21.4	22.6
140	13.S	14.7	15.6	16.6	17.5	IS.4	19.3	20.2	21.1	22.0

Table 1.7 Selected Crude Oils Showing the Differences in API Gravity and Sulfur Content Within a Country.

Country	Crude oil	API	Sulfur % w/w
Abu Dhabi (UAE)	Abu Al Bu Khoosh	31.6	2.00
Abu Dhabi (UAE)	Murban	40.5	0.78
Angola	Cabinda	31.7	0.17
Angola	Palanca	40.1	0.11
Australia	Barrow Island	37.3	0.05
Australia	Griffin	55.0	0.03
Brazil	Garoupa	30.0	0.68
Brazil	Sergipano Platforma	38.4	0.19
Brunei	Champion Export	23.9	0.12
Brunei	Seria	40.5	0.06
Cameroon	Lokele	20.7	0.46
Cameroon	Kole Marine	32.6	0.33
Canada (Alberta)	Wainwright-Kinsella	23.1	2.58
Canada (Alberta)	Rainbow	40.7	0.50
China	Shengli	24.2	1.00
China	Nanhai Light	40.6	0.06
Dubai (UAE)	Fateh	31.1	2.00
Dubai (UAE)	Margham Light	50.3	0.04
Egypt	Ras Gharib	21.5	3.64
Egypt	Gulf of Suez	31.9	1.52
Gabon	Gamba	31.4	0.09
Gabon	Rabi-Kounga	33.5	0.07
Indonesia	Bima	21.1	0.25
Indonesia	Kakap	51.5	0.05
Iran	Aboozar (Ardeshir)	26.9	2.48
Iran	Rostam	35.9	1.55
Iraq	Basrah Heavy	24.7	3.50
Iraq	Basrah Light	33.7	1.95
Libya	Buri	26.2	1.76
Libya	Bu Attifel	43.3	0.04

(Continued)

Table 1.7 Cont.

Country	Crude oil	API	Sulfur % w/w
Malaysia	Bintulu	28.1	0.08
Malaysia	Dulang	39.0	0.12
Mexico	Maya	22.2	3.30
Mexico	Olmeca	39.8	0.80
Nigeria	Bonny Medium	25.2	0.23
Nigeria	Brass River	42.8	0.06
North Sea (Norway)	Emerald	22.0	0.75
North Sea (UK)	Innes	45.7	0.13
Qatar	Qatar Marine	36.0	1.42
Qatar	Dukhan (Qatar Land)	40.9	1.27
Saudi Arabia	Arab Heavy (Safaniya)	27.4	2.80
Saudi Arabia	Arab Extra Light (Berri)	37.2	1.15
USA (California)	Huntington Beach	20.7	1.38
USA (Michigan)	Lakehead Sweet	47.0	0.31
Venezeula	Leona	24.4	1.51
Venezuela	Oficina	33.3	0.78

Table 1.8 API Gravity and Sulfur Content of Selected Heavy Oils and Tar Sand Bitumen.

Country	Crude oil	API	Sulfur % w/w
Canada (Alberta)	Athabasca	8.0	4.8
Canada (Alberta)	Cold Lake	13.2	4.11
Canada (Alberta)	Lloydminster	16.0	2.60
Canada (Alberta)	Wabasca	19.6	3.90
Chad	Bolobo	16.8	0.14
Chad	Kome	18.5	0.20
China	Qinhuangdao	16.0	0.26
China	Zhao Dong	18.4	0.25
Colombia	Castilla	13.3	0.22
Colombia	Chichimene	19.8	1.12

(Continued)

Table 1.8 Cont.

Country	Crude oil	API	Sulfur % w/w
Ecuador	Ecuador Heavy	18.2	2.23
Ecuador	Napo	19.2	1.98
USA (California)	Midway Sunset	11.0	1.55
USA (California)	Wilmington	18.6	1.59
Venezuela	Boscan	10.1	5.50
Venezuela	Tremblador	19.0	0.80

1.07 Barrel – Conversion to other Units.

	To convert to:				
Crude oil	Tonnes (metric)	Liters × 1000	Barrels	US gallons	Tonnes/ year
From	Multiply by				
Barrels	0.1364	0.159	1	42	–
Barrels per day	–	–	–	–	49.8
Liters (× 1000)	0.8581	1	6.2898	264.17	–
Tonnes (metric)	1	1.165	7.33	307.86	–
US gallons	0.00325	0.0038	0.0238	1	–

1.08 Bernoulli's Principle

Bernoulli's principle states that an increase in the speed of a fluid occurs simultaneously with a decrease in pressure or a decrease in the potential energy of the fluid. The principle can be applied to various types of fluid flow, resulting in various forms of the Bernoulli equation. A common form of Bernoulli's equation, valid at any arbitrary point along a streamline is:

$$v^2/2 + gz + p/\rho = \text{constant}$$

In this equation, v is the fluid flow speed at a point on a streamline, *g* is the acceleration due to gravity, *z* is the elevation of the point above a reference plane, with the positive *z*-direction pointing upward – so in the direction opposite to the gravitational acceleration, *p* is the pressure at the chosen point, and ρ is the density of the fluid at all points in the fluid. The constant on the right-hand side of the equation depends only on the streamline chosen, whereas v, z, and p depend on the particular point on that streamline.

In many applications of Bernoulli's equation, the change in the ρ g z term along the streamline is so small compared with the other terms that it can be ignored. This allows the above equation to be presented in a simplified form in which p_0 is the total pressure and q is the dynamic pressure. Thus:

$$p + q = p_o$$

Static pressure + dynamic pressure = total pressure.

Every point in a steadily flowing fluid, regardless of the fluid speed at that point, has its own unique static pressure, p, and dynamic pressure, q. Their sum $(p + q)$ is defined to be the total pressure, p_0. The significance of Bernoulli's principle can now be summarized as *total pressure is constant along a streamline.*

1.09 Brine

Brine is an aqueous solution of salts that occur with gas and crude oil; seawater and saltwater are also known as brine. At 15.5°C (60°F) saturated sodium chloride brine is 26.4% sodium chloride by weight (100 degree SAL). At 0°C (32°F) brine can only hold 26.3% salt. Brine is at the high end of the water salinity scale (Table 1.9). Brine is corrosive to metal and there must be periodic inspection of pipelines and other metals systems with which brine comes into contact.

1.10 Bubble Point and Bubble Point Pressure

The bubble point is the temperature at which incipient vaporization of a liquid in a liquid mixture occurs, corresponding with the equilibrium point of 0 per cent vaporization or 100 per cent condensation. At a given

Table 1.9 Water Salinity Based on Dissolved Salts (parts per thousand).

Fresh water	Brackish water	Saline water	Brine
<0.5	0.5–30	30–50	>50

temperature, when the pressure decreases and below the bubble point curve, gas will be emitted from the liquid phase to the two-phase region.

At the bubble point, the following relationship holds:

$$\sum_{i=1}^{N_c} y_i = \sum_{i=1}^{N_c} K_i x_i = 1$$

K is the *distribution coefficient* (*K factor*) which is the ratio of mole fraction in the vapor phase (y_{ie}) to the mole fraction in the liquid phase (x_{ie}) at equilibrium. When Raoult's law and Dalton's law hold for the mixture, the K factor is defined as the ratio of the vapor pressure to the total pressure of the system:

$$K_i = y_{ie}/x_{ie}$$

Given either of x_i or y_i and either the temperature or pressure of a two-component system, calculations can be performed to determine the unknown information.

The *bubble point pressure* (P_b) is the pressure at which saturation will occur in the liquid phase (for a given temperature) and is the point at which vapor (bubble) first starts to come out of the liquid (due to pressure depletion). The bubble-point pressure p_b of a hydrocarbon system is the highest pressure at which a bubble of gas is first liberated from the oil. This important property can be measured experimentally for a crude oil system by conducting a constant-composition expansion test.

The bubble point temperature is usually lower than the dew point temperature for a given mixture at a given pressure (Figure 1.1). Since the vapor above a liquid will probably have a different composition to the liquid, the bubble point (along with the *dew point*) data at different compositions are useful data when designing distillation systems and for constructing phase diagrams as a means of studying phase relationships. As pressures are reduced below the bubble point, the relative volume of the gas phase increases. For pressures above the bubble point, a crude oil is said to undersaturated. At or below the bubble point, the crude is saturated.

In the absence of the experimentally measured bubble-point pressure, it is necessary to make an estimate of this crude oil property from the readily available measured producing parameters – these correlations assume that the bubble-point pressure is a strong function of gas solubility R_s, gas gravity γ_g, oil gravity in °API, and temperature T:

$$p_b = f(R_s, \gamma_g, °API, T)$$

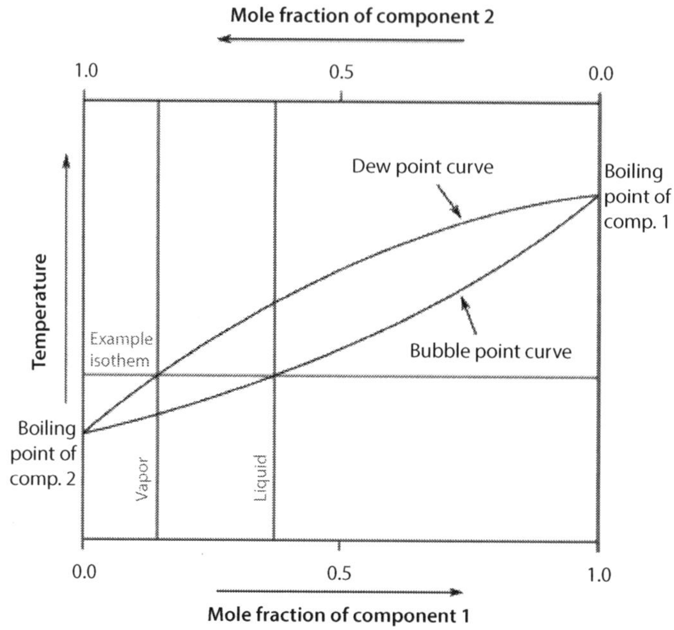

Figure 1.1 Relationship of Bubble Point to Dew Point.

1.11 Buoyancy, Buoyed Weight, and Buoyancy Factor

Buoyancy is the upward force exerted by a fluid that opposes the weight of an immersed object. In a column of fluid, pressure increases with depth because of the weight of the overlying fluid. Thus, the pressure at the bottom of a column of fluid is greater than at the top of the column and the pressure at the bottom of an object submerged in a fluid is greater than at the top of the object. This pressure difference results in a net upwards force on the object and the magnitude of that force exerted is proportional to that pressure difference, and is equivalent to the weight of the fluid that would otherwise occupy the volume of the object, i.e. the displaced fluid. Thus:

$$\text{Buoyancy} = (\text{weight of material in air}) / (\text{density of material}) \times \text{density}$$

$$\text{Buoyancy weight} = (\text{density of material in air} - \text{fluid density}) / (\text{density of material}) \times (\text{weight of material in air})$$

$$\text{Buoyancy factor} = (\text{density of material in air} - \text{fluid density})/(\text{density of material}) = (\rho_s - \rho_m)/\rho_s = 1 - \rho_m/\rho_s$$

ρ_s is the density of the steel/material, and *r* m is the density of the fluid/mud. When the inside and outside fluid densities are different, the buoyancy factor can be given as:

$$\text{Buoyancy factor} = [A_o(1 - \rho_o/\rho_s) - A_i(1 - \rho_i/\rho_s)]/A_o - A_i$$

A_o is the external area of the component, and A_i is the internal area of the component.

1.12 Capacity

The capacity of a pipe, the annular capacity, and the annular volume can be calculated using the following equations. The linear capacity of the pipe is:

$$C_i = A_i/808.5 \text{ bbl/ft}$$

A_i is a cross-sectional area of the inside pipe in square inches and is equal to $0.7854 \times Di^2$ and D_i is the inside diameter of the pipe in inches.

The volume capacity is:

$$V = C_i \times L\ bbl$$

L = the length of the pipe, in feet.

The annular linear capacity against the pipe is:

$$C_o = A_o/808.5 \text{ bbl/ft)}$$

A_o is the cross-sectional area of the annulus in square inches and is obtained from the relationship:

$$0.7854 \times (D_h^2 - D_o^2)$$

D_o = the outside side diameter of the pipe, in inches, and D_h = the diameter of the hole or the inside diameter of the casing against the pipe, in inches. Thus, the annular volume capacity is:

$$V = C_o \times L \text{ bbl}$$

1.12.1 Hole (Pipe, Tubing) Capacity (in barrels per one linear foot, bbl/ft)

These equations are applicable to calculating internal volume and displacement for hole, pipe, or tubing using the inside diameter in inches.

$$C = ID^2/1029.4$$

C – capacity. Bbl/ft
ID – inside diameter of hole, pipe, tubing, inches, 1029.4 = conversion factor, inches²-ft/bbl

To determine the total volume of a hole, pipe, or tubing, multiply the capacity by the length of the hole or pipe in feet:

$$V_{tot} = C \times h$$

V_{tot} is the total volume of hole or pipe, bbl, C is the capacity of hole or pipe, bbl/ft, H is the length of hole or pipe, ft.

1.12.2 Annular Capacity

The values derived using the following equations are applicable to any combination of hole, casing, or liner on the outside and tubing or drill pipe on the inside. Thus:

$$C_{an} = (ID^2 - OD^2)/1029.4$$

C_{an} is capacity of annular space per lineal foot, bbl/ft, ID inside (casing, liner) diameter, inches, OD is the outside (work string, tubing) diameter in inches, 1029.4 is the conversion factor.

1.12.3 Annular Volume (volume between casing and tubing, bbl)

$$V_{an} = C_{an} \times h$$

V_{an} is the total volume of annulus with piping/tubing in well, bbl, C_{an} is the capacity of annulus, bbl/ft, h is the length of annulus, ft.

Related to capacity, the velocity of the fluid is given by the relationship:

$$Vel = Q/C$$

Vel is the velocity, ft/min, Q is the flow rate, bbl/min, C is the capacity of hole, pipe, annulus, bbl/ft.

1.13 Capillary Number

The capillary number (N_c) is the ratio of viscous forces to capillary forces, and equal to viscosity times velocity divided by interfacial tension. A common experimental observation is a relationship between residual oil saturation (S_{or}) and local capillary number (N_c). This relationship is called a capillary desaturation curve (CDC).

The capillary number reflects the balance between viscous and capillary forces at the pore scale; viscous forces dominate at high capillary numbers while capillary forces dominate at low capillary numbers. The capillary number at the pore scale is defined as:

$$N_c = (\mu v)/(\gamma \cos\theta)$$

In the equation, μ is the water viscosity, ν is the linear advance rate, γ is the oil-water interfacial tension and θ is the contact angle.

If the viscous forces acting on the trapped oil exceed the capillary retaining forces, residual oil can be mobilized.

1.14 Capillary Pressure

The capillary pressure between adjacent oil and water phases, P_{cow}, can be related to the principal radii of curvature R_1 and R_2 of the shared interface and the interfacial tension σ_{ow} for the oil/water interface:

$$P_{cow} = p_o - p_w = \sigma_{ow}(1/R_1 + 1/R_2)$$

In this equation,

p_o = pressure in the oil phase, m/Lt ^{2}, psi
p_w = pressure in the water phase, m/Lt ^{2}, psi
P_{co} = capillary pressure between oil and water phases, m/Lt^2, psi
R_1, R_2 = principal radii of curvature, L
σ_{ow} = oil/water interfacial tension, m/t ^{2}, dyne/cm

The displacement of one fluid by another in the pores of a porous medium is either aided or opposed by the surface forces of capillary pressure. As a consequence, in order to maintain a porous medium partially-saturated with non-wetting fluid and while the medium is also exposed to wetting fluid, it is necessary to maintain the pressure of the non-wetting fluid at a value greater than that in the wetting fluid. Also, denoting the pressure in the wetting fluid by P_w and that in the non-wetting fluid by P_{nw}, the capillary pressure can be expressed as:

$$P_c = P_{nw} - P_w$$

The pressure excess in the non-wetting fluid is the capillary pressure, and this quantity is a function of saturation. In addition, there are three types of capillary pressure: (1) water-oil capillary pressure, P_{cwo}, (2) gas-oil capillary pressure, P_{cgo}, and gas-water capillary pressure (P_{cgw}). Applying the mathematical definition of the capillary pressure, the three types of the capillary pressure can be written as:

$$P_{cwo} = P_o - P_w$$

$$P_{cgo} = P_g - P_o$$

$$P_{cgw} = P_g - P_w$$

1.15 Cementation Value

Cementation refers to the event in a sediment where new minerals stick the grains together and is relevant to the ease of crude oil flow from the reservoir rock. The cementation value (cementation factor, also the cementation exponent, *m*) varies from approximately 1.3 to 2.6 (Table 1.10) and

Table 1.10 Lithology and Cementation Values.

Lithology	**Cementation value**
Unconsolidated rocks (loose sands limestones)	1.3
Very slightly cemented	1.4–1.5
Slightly cemented (sands with >20% porosity	1.6–1.7
Moderately cemented (consolidated <15%)	1.8–1.9
Highly cemented (quartzite, limestone, dolomite)	2.0–2.2

is dependent on (or an indicator of) the rock lithology, especially (1) the shape, type, and size of grains, (2) the shape and size of pores and pore throats, and (3) the size and number of dead-end (or cul-de-sac) pores. The dependence of the cementation factor on the degree of cementation is not as strong as its dependence on the shape of grains and pores.

1.16 Composite Materials

For longitudinal directional ply and longitudinal tension, the modulus is:

$$E = V_m E_m + V_f E_f$$

V_m is the volume fraction of the matrix, E_m is the elastic modulus of the base pipe, V_f is the volume fraction of the fiber attachment, and E_f is the elastic modulus of the rubber attachment. Also, $V_m + V_f = 1$

Example calculation
Estimate the modulus of the composite shaft with 25% of the total volume with fibers. Assume that the modulus of elasticity for the fiber is 50×10^6 psi, the modulus of elasticity for the matrix is 600 psi, and the load is applied longitudinally as well as perpendicular to the fibers.

Solution:
When the load is applied longitudinally to the fibers:

$$E = V_m E_m + V_f E_f = 500 \times 0.75 + 25{,}000{,}000 \times 0.25 = 6{,}250{,}450$$

When the load is applied perpendicular to the fibers:

$$1/E = {}_{V}m/{}_{E}m + {}_{V}f/{}_{E}f = 0.25/500 + 0.75/25{,}000{,}000 = 0.00125$$

Thus, E = 800 psi.

1.17 Compressibility

The compressibility factor Z is a dimensionless factor independent of the quantity of gas and determined by the character of the gas, the temperature, and pressure (see Table 1.11 for the meaning of the symbols):

$$Z = PV/NRT + MPV/mRT$$

Table 1.11 Symbols used in Determining the Compressibility Factor.

	Field units	**SI units**
P = absolute pressure	psia	kPa
V = volume	ft^3	m^3
n = moles	m/M	m/M
m = mass	lb	kg
M = molecular weight	lb/lb mole	kg/kmole
T = absolute temperature	°R	K
P = density	slug/ft^3	kg.m^3

A knowledge of the compressibility factor means that the density, ρ, is also known from the relationship:

$$\rho = PM/ZRT$$

The isothermal gas compressibility, which is given the symbol c_g, is a useful concept which will be used extensively in determining the compressible properties of the reservoir. The isothermal compressibility is also called the bulk modulus of elasticity. Gas usually is the most compressible medium in the reservoir. However, care should be taken so that it not be confused with the gas deviation factor, z, which is sometimes called the super-compressibility factor.

The isothermal gas compressibility is defined as:

$$C_g = (1/V_g)(\delta V_g/\delta p)_T$$

An expression in terms of z and p for the compressibility can be derived from the ideal gas law:

$$(\delta V_g/\delta p)_T = (nRT/p)(\ \delta z/\delta p)_T = znRT/p^2 =$$
$$(znR'T/p)1/z(dz/dp) - znR'T/p) \times 1/p$$

From the real gas equation of state:

$$1/V_g = p/(znRT)$$

Hence:

$$1/V_g(\delta V_g/\delta p) = i/z(dz/dp) - 1/p$$

Thus:

$$C_g = 1/p - 1/z(\delta z/\delta p)_T$$

For gases at low pressures the second term is small and the compressibility can be approximated by $c_g \approx 1/p$. Equation 2 is not particularly convenient for determining the gas compressibility because z is not actually a function of p but of p_r. However, equation 2 can be made convenient in terms of a dimension-less pseudo-reduced gas compressibility defined as:

$$c_r = c_g p_{pc}$$

Multiplying equation 2 through by the pseudo-critical pressure,

$$c_r = c_g - p_{pc} = 1/p_r = 1/z(\delta z/\delta p_r)_{Tr}$$

The expression for calculating the pseudo-reduced compressibility is:

$$c_r = (1/p_r) - 0.27/z^2 T_r[(\delta z/\delta p_r)_{Tr})/1 + (\rho_r/z)(\delta z/\delta p_r)_{Tr}]$$

There is also a close relationship between the formation volume factor of gas and the isothermal gas compressibility. It can be easily shown that:

$$Cg = (1/B_g)/\rho B_g/\delta p)_T$$

1.18 Darcy's Law

For laminar single fluid flows in straight ducts, the flow resistance or pressure drop is proportional to the flow rate. The same relationship holds for flow in curved ducts when the flow velocity is very small. This unique relationship between the flow velocity and the pressure drop can be generalized to flow through porous media as well,

$$dp/dx - \rho g_x \sim -u$$

In this equation, u is the superficial fluid flow velocity (or discharge rate per unit cross-sectional area), p is the fluid pressure, x is the linear coordinate in the flow direction, ρ is the fluid density and g_x is the gravity in the direction of flow. This limiting flow behavior is the basis for macroscopic modeling flow through packed beds; the proportionality constant can be determined if a flow geometry is defined.

A one-dimensional empirical model continuum, for saturated single fluid flow in porous media was based on the proportionality and Darcy's law can also be expressed as:

$$u = k/\mu(\delta p/\delta x - \rho g_x)$$

Where k is the permeability of the porous medium which is assumed to be constant in applications and μ is the dynamic viscosity of the fluid. Darcy's law has been generalized to be used for multi-dimensional single phase and multiphase flows. Here, multiphase flows specifically mean immiscible multiphase flows. For miscible systems, one can effectively treat them as single phase flows.

For single phase flows, Darcy's law lacks both the flow diffusion effects and the inertial effects. Therefore, the utility of Darcy's law is restrictive and validation of the modeling results is often necessary. Remedies of these defects have been adjusted by addition of a diffusion term to the Darcy's law:

$$Vp - \rho g = -(\mu/k)v + \mu V^2 v$$

$\breve{\mu}$ is an effective viscosity and v is the superficial fluid flow velocity field. In general, the effective viscosity $\breve{\mu}$ is proportional to the fluid viscosity μ and is affected by the type of porous media. For simplicity and convenience, the effective viscosity is usually taken to be identical to the fluid viscosity.

1.19 Dew Point Temperature and Pressure

The *dew point pressure* (P_d) is the pressure at which the first condensate liquid comes out of solution in a gas condensate. Thus, the dew point curve is the curve that separates the pure gas phase from the two-phase region and represents the pressure and temperature at which the first liquid droplet is formed out of the gas phase.

The *dew point temperature* is a measure of how much water vapor there is in a gas. Water has the property of being able to exist as a liquid, solid, or gas under a wide range of conditions. To understand the behavior of water vapor, it is first useful to consider the general behavior of gases. In any mixture of gases, the total pressure of the gas is the sum of the partial pressures of the component gases. This is Dalton's law and it is represented as follows:

$$P_{total} = P_1 + P_2 + P_3 \ldots \text{ etc.}$$

The quantity of any gas in a mixture can be expressed as a pressure.

1.20 Displacement

The open-ended displacement volume of a pipe is calculated as follows:

$$V_o = (0.7854(D_o^2 - D_i^2)/808.5 \text{ bbl/ft}$$

$$\text{Displacement volume} = V_o \times L \text{ bbl}$$

The close-ended displacement volume of the pipe is calculated from:

$$V_c = 0.7854(D_o^2)/808.5 \text{ bbl ft}$$

$$\text{Displacement volume} = V_c \times L \text{ bbl}$$

Example calculation:
Calculate the drill pipe capacity, open-end displacement, closed-end displacement, annular volume, and total volume for the following condition: 5,000 feet of 5" drill pipe with an inside diameter of 4.276" inside a hole of 8.".

Solution:
The linear capacity of pipe, C_i, is calculated as:

$$C_i = A_i/808.5 = (0.7854 \times D_i^2)/808.5 =$$
$$(0.7854 \times 4.276_2)/808.5 = 0.017762 \text{ bbl/ft}$$

Thus, the pipe volume capacity is: 0.017762 × 5000 = 0.006524 bbl/ft

The open-end displacement of pipe, V_o, is:

$$V_o = [0.7854(D_o^2 - D_i^2)].808.5 =$$
$$[0.7854(5^2 - 4.276^2)]/808.5 = 0.006524 \text{ bbl/ft}$$

The close-end displacement volume of the pipe, V_c, is:

$$V_c = [0.7854(D_o^2)]/808.5 = [0.7854(5^2)]/808.5 = 0.024286 \text{ bbl/ft}$$

The annular volume of the pipe, V, is:

$$V = C_o \times L = A_o/808.5 \times L = 0.7854/808.5 \times (D_h^2 - D_o^2) \times L = 0.7854/808.5 \times (8.5^2 - 5^2) \times 5000 = 229.5 \text{ bbl}$$

1.21 Effective Weight

The effective weight per unit length can be calculated using the following relation in which the weight per foot in drilling mud is the weight per foot in air minus the weight per foot of the displaced drilling mud:

$$w_b = w_s + \rho_i A_i - \rho_o A_o$$

$$A_o = \pi/4(0.95 \times D_o^2 + 0.05 \times D_{oj}^2$$

$$A_i = \pi/4(0.95 \times D_i^2 + 0.05 \times D_{ij}^2)$$

Without tool joints:

$$A_i = 0.7854 \times D_i^2$$

$$A_o = 0.7854 \times D_o^2$$

Thus:

$$w_b = w_s + \rho_i A_i - \rho_o A_o$$

In the above equation, unit weight of the steel can be given as:

$$w_s = \rho_s A_s$$

If the inside and outside fluid densities are the same, thus:

$$w_b = A_s(\rho_s - \rho_o) = A_s \rho_s (1 - \rho_o/\rho_s) = w_s(1 - \rho_o/\rho_s)$$

In this equation, D_o is the outside diameter of the component body, D_{oj} is the outside diameter of the tool joint, D_i is the inside diameter of the component body, D_{ij} is the inside diameter of the tool joint, A_s is the cross-sectional area of the steel/material, ρ_o is the annular mud weight at component depth in the wellbore, ρ_i is the internal mud weight at component depth inside the component, and ρ_s is the density of the steel/material.

Example calculation
Calculate the buoyancy factor and buoyed weight of 6,000 ft of 6 5/8" 27.7 ppf E grade drill pipe in mud of density 10 ppg.

Solution:
Using a steel density of 65.4 ppg,

$$\text{Buoyancy factor} = (1 - \rho_m/\rho_s) = (1 - 10/65.4) = 0.847$$

$$\text{Douyed weight} = 0.847 \times 27.7 \times 6000 = 140771.4 \text{ lbf} = 140 \text{ kips}$$

1.22 Flow Through Permeable Media

1.22.1 Productivity Index

The productivity index, *J*, of an oil well is the ratio of the stabilized rate, *q*, to the pressure drawdown required to sustain that rate (see Table 1.12 for a definition of the various symbols). For flow from a well centered in a circular drainage area, the productivity index can be related to formation and fluid properties:

$$J = q/p - p_{wf} = kh/(141.2B\mu)/[\ln(r_e/r_w) - 3/4 + s$$

The productivity index can also be expressed for general drainage-area geometry as:

$$J = q/p - p_{wf} = (0.00708kh)/B\mu[1/2\ln(10.06A/C_A r_w^2) - 3/4 + s]$$

1.22.2 Steady-State Flow

Pseudo-steady-state flow describes production from a closed drainage area (one with no-flow outer boundaries, either permanent and caused by zero-permeability rock or temporary and caused by production from

Table 1.12 Definition of the various symbols.

a	=	$\frac{1.422\times10^{6}}{k_g h}\left[1.151\log\left(\frac{10.06A}{C_A r_w^2}\right)\right]-\frac{3}{4}+s$, stabilized deliverability coefficient, psia^2-cp/MMscf-D
a	=	total length of reservoir perpendicular to wellbore, feet
a_h	=	length of reservoir perpendicular to horizontal well, feet
a_f	=	$(L_f^2 + b_f^2)^{1/2}$, depth of investigation along major axis in fractured well, feet
a_t	=	$(L_f^2 + b_f^2)$, transient deliverability coefficient, psia^2-cp/MMscf-D
a_H	=	total width of reservoir perpendicular to the wellbore, feet
$a_H{}'$	=	modified total width of reservoir perpendicular to the wellbore, feet
A	=	drainage area, sq feet
A	=	$\pi a_f b_f$, area of investigation in fractured well, feet^2
A_f	=	cross-sectional area perpendicular to flow, sq feet
A_{wb}	=	wellbore area, sq feet
b	=	1.422×10^6TD)/kgh (gas flow equation)
b_f	=	$0.02878(kt/\varphi\mu c_t)^{1/2}$, depth of investigation of along minor axis in fractured well, feet
b_B	=	intercept of Cartesian plot of bilinear flow data, psi
b_H	=	length in direction parallel to wellbore, feet
$b_H{}'$	=	modified length in direction parallel to wellbore, feet
b_L	=	intercept of Cartesian plot of linear flow data, psi
b_V	=	intercept of Cartesian plot of data during volumetric behavior, psi
B	=	formation volume factor, reservoir volume/surface volume
B_g	=	gas formation volume factor, RB/STB
B_{gi}	=	gas formation volume factor evaluated at p_i, RB/Mscf
B_o	=	oil formation volume factor, RB/Mscf
B_w	=	water formation volume factor, RB/STB
$\bar{B}_g$	=	gas formation volume factor evaluated at average drainage area pressure, RB/Mscf
B_{ND}	=	1,422 μ z TD/kh, non-Darcy flow coefficient

(*Continued*)

Table 1.12 Cont.

c	=	compressibility, psi^{-1}
c_f	=	formation compressibility, psi^{-1}
c_g	=	gas compressibility, psi^{-1}
c_o	=	oil compressibility, psi^{-1}
c_t	=	$S_o c_o + S_w c_w + S_g c_g + c_f$ = total compressibility, psi^{-1}
c_w	=	water compressibility, psi^{-1}
$\overline{c}_t$	=	total compressibility evaluated at average drainage area pressure, psi^{-1}
c_{tf}	=	total compressibility of pore space and fluids in fracture porosity, psi^{-1}
c_{tm}	=	total compressibility of pore space and fluids in matrix porosity, psi^{-1}
c_{wb}	=	compressibility of fluid in wellbore, psi^{-1}
C	=	performance coefficient in gas-well deliverability equation, or wellbore storage coefficient, bbl/psi
C_A	=	shape factor or constant
C_D	=	0.8936 $C/\phi c_t h r_w^2$, dimensionless wellbore storage coefficient
$(C_D e^{2s})_f$	=	type-curve parameter value for the formation
$(C_D e^{2s})_{f+m}$	=	type-curve parameter value for the formation plus the matrix
C_{LfD}	=	0.8936 $C / \phi c_t h L_f^2$, dimensionless wellbore storage coefficient in fractured well
C_r	=	$w_f k_f / \pi k L_f$, fracture conductivity, dimensionless
d_x	=	shortest distance between horizontal well and *x* boundary, feet
d_y	=	shortest distance between tip of horizontal well and *y* boundary, feet
d_z	=	shortest distance between horizontal well and *z* boundary, feet
D_x	=	longest distance between horizontal well and × boundary, feet
D_y	=	longest distance between tip of horizontal well and y boundary, feet
D_z	=	longest distance between horizontal well and z boundary, feet
D	=	non-Darcy flow constant, D/Mscf
e^{-bt}	=	exponential decline with a constant *b* and elapsed time, *t*
E_f	=	flow efficiency, dimensionless

(*Continued*)

Table 1.12 Cont.

$Ei(-x)$	=	$\int_x^\infty (e^{-u}/u)du$, the exponential integral
$F(u)$	=	function used in horizontal well analysis
F_{CD}	=	$w_f k_f / kL_f$, fracture conductivity, dimensionless
g	=	acceleration due to gravity, feet/sec^2
g_c	=	gravitational units conversion factor, 32.17 (lbm/feet)(lbf-s^2)
h	=	net formation thickness, feet
h_D	=	$(h/r_w)(k_h/k_v)^{1/2}$, dimensionless
h_f	=	fracture height, feet
h_m	=	thickness of matrix, feet
h_p	=	perforated interval thickness, feet
h_{pD}	=	h_p/h_t
h_t	=	total formation thickness, feet
h_1	=	distance from top of formation to top of perforations, feet
h_{1D}	=	h_1/h_t
HTR_{avg}	=	HTR at average drainage area pressure
J	=	productivity index, STB/D, psi
J_{actual}	=	actual well productivity index, STB/D-psi
J_{ideal}	=	ideal productivity index ($s = 0$), STB/D-psi
k	=	matrix permeability, md
$\bar{k}$	=	average permeability, md
k_f	=	permeability of the proppant in the fracture, md
k_{fs}	=	permeability near the wellbore, md
k_g	=	permeability to gas, md
k_{gp}	=	permeability of the gravel in the gravel pack, md
k_h	=	horizontal permeability, md
k_m	=	matrix permeability, md
k_o	=	permeability to oil, md
k_r	=	permeability in horizontal radial direction, md
k_s	=	permeability of altered zone, md
k_w	=	permeability to water, md
k_x	=	permeability in x-direction, md

(Continued)

Table 1.12 Cont.

k_y	=	permeability in *y*-direction, md
k_z	=	permeability in *z*-direction, md
L	=	distance from well to no-flow boundary, feet
L_d	=	drilled length of horizontal well, feet
L_f	=	fracture half length, feet
L_g	=	length of flow path through gravel pack, feet
L_m	=	length of matrix, feet
L_p	=	length of perforation tunnel, feet
L_s	=	length of damaged zone in fracture, feet
L_w	=	completed length of horizontal well, feet
L_x	=	distance from boundary, feet
m	=	162.2 $qB\mu/kh$ = slope of middle-time line, psi/cycle
m_B	=	slope of bilinear flow graph, psi/hr$^{1/4}$
m_L	=	slope of linear flow graph, psi/hr$^{1/2}$
m_s	=	$\frac{2456\sqrt{\phi\mu c_t qB\mu}}{k_{sp}^{3/2}}$, slope of spherical flow plot, psi-hr$^{1/2}$
m_V	=	slope of volumetric flow graph, psi/hr
m_{hrf}	=	slope of semilog plot for hemiradial flow, psi/log cycle
m_{elf}	=	slope of square-root-of-time plot for early linear flow, psi/$\sqrt{hr}$
m_{erf}	=	slope of semilog plot of early radial flow, psi/log cycle
m_{llf}	=	slope of square-root-of-time plot for late linear flow, psi/$\sqrt{hr}$
m_{prf}	=	slope of semilog plot for pseudoradial flow, psi/log cycle
M	=	Molecular weight of gas
MTR	=	middle-time region
n	=	inverse slope of the line on a log-log plot of the change in pressure squared or pseudo pressure vs. gas flow rate
p	=	pressure, psi
p_{avg}	=	average pressure, psi
p_b	=	base (atmospheric) pressure, psia
p_0	=	arbitrary reference or base pressure, psi
$\overline{p}$	=	volumetric average or static drainage-area pressure, psi

(*Continued*)

Table 1.12 Cont.

p_a	=	adjusted or normalized pseudo pressure, $(\mu z/p)p_p$, psia
p_{awf}	=	adjusted flowing bottomhole pressure, psia
p_{aws}	=	adjusted shut-in bottomhole pressure, psia
p_f	=	formation pressure, psi
p_i	=	original reservoir pressure, psi
p_m	=	matrix pressure, psi
p_p	=	pseudopressure, $psia^2/cp$
p_s	=	stabilized shut-in BHP measured just before start of a deliverability test, psia
p_{sc}	=	standard-condition pressure, psia
p_t	=	surface pressure in tubing, psi
p_w	=	BHP in wellbore, psi
p_{wf}	=	flowing BHP, psi
p_{ws}	=	shut-in BHP, psi
p_{xy}	=	parameter in horizontal well analysis equations
p_{xyz}	=	parameter in horizontal well analysis equations
p_y	=	parameter in horizontal well analysis equations
p_{1hr}	=	pressure at 1-hour shut-in (flow) time on MTR line or its extrapolation, psi
p'	=	pressure derivative
p^*	=	MTR pressure trend extrapolated to infinite shut-in time, psi
p_D	=	$0.00708\, kh(p_i - p)/qB\mu$, dimensionless pressure as defined for constant-rate production
p_{MBHD}	=	Matthews-Brons-Hazebroek pressure, dimensionless
$(p_D)_{MP}$	=	dimensionless pressure at match point
q	=	flow rate at surface, STB/D
q_{AOF}	=	absolute-open-flow potential, MMscf/D
q_g	=	gas flow rate, Mscf/D
q_o	=	water flow rate, STB/D
q_{Rt}	=	total flow rate at reservoir conditions, RB/D
q_{sf}	=	flow rate at formation (sand) face, STB/D
q_w	=	water flow rate, STB/D
r	=	distance from the center of wellbore, feet
r_a	=	radius of altered zone (skin effect), feet

(*Continued*)

Table 1.12 Cont.

r_d	=	effective drainage radius, feet
r_{dp}	=	radius of damage zone around perforation tunnel, feet
r_e	=	external drainage radius, feet
r_i	=	radius of investigation, feet
r_p	=	radius of perforation tunnel, feet
r_s	=	outer radius of the altered zone, feet
r_{sp}	=	radius of source or inner boundary of spherical flow pattern, feet
r_w	=	wellbore radius, feet
r_{wa}	=	apparent or effective wellbore radius, feet
r_D	=	r/r_w, dimensionless radius
R_s	=	dissolved GOR, scf/STB
s	=	skin factor, dimensionless
s_a	=	skin caused by alteration of permeability around wellbore, dimensionless
s_c	=	convergence skin, dimensionless
s_d	=	skin caused by formation damage, dimensionless
s_e	=	skin caused by eccentric effects, dimensionless
s_{dp}	=	perforation damage skin, dimensionless
s_f	=	skin of hydraulically fractured well, dimensionless
s_{gp}	=	skin factor from to Darcy flow through gravel pack, dimensionless
s_{min}	=	minimum skin factor, dimensionless
s_p	=	skin resulting from an incompletely perforated interval, dimensionless
s_t	=	total skin, dimensionless
s_θ	=	skin factor resulting from well inclination, dimensionless
s'	=	$s + Dq$ = apparent skin factor, dimensionless
S_g	=	gas saturation, fraction of pore volume
S_o	=	oil saturation, fraction of pore volume
S_w	=	water saturation, fraction of pore volume
t	=	elapsed time, hours
t_a	=	$\mu c_t t_{ap}$, adjusted or normalized pseudo time, hours
t_{ap}	=	pseudo time, hours

(Continued)

Table 1.12 Cont.

t_{bD}	=	dimensionless time in linear flow, hours
t_D	=	0.0002637 $kt/\phi\mu c_t r_w^{\ 2}$, dimensionless time
t_{DA}	=	0.0002637 $kt/\phi\mu c_t A$ = dimensionless time based on drainage area, A
t_{eqB}	=	equivalent time for bilinear flow, hours
t_e	=	equivalent time, hours
t_{LfD}	=	0.0002637 $kt/\phi\mu e_t L_f^2$, dimensionless time for fractured wells
t_p	=	pseudo producing time, hours
t_{pD}	=	pseudo producing time, dimensionless
t_{prf}	=	time required to reach the pseudoradial flow regime, hours
t_{Eelf}	=	end of early linear flow, t, hours
t_{Eerf}	=	end of early radial flow, t, hours
t_{Elf}	=	end of linear flow, hours
t_{Ellf}	=	time to end of late linear flow regime, hours
t_{Ehrf}	=	end of hemiradial flow, hours
t_{Erf}	=	end of early radial flow, hours
t_{Eprf}	=	end of pseudoradial flow, hours
t_p	=	constant-rate production period, t, hours
t_{pAD}	=	dimensionless producing time, hours
t_{pss}	=	time required to reach pseudo steady state, hours
t_{Self}	=	start of early linear flow, hours
t_{Sllf}	=	start of late linear flow, hours
t_{Shrf}	=	start of hemiradial flow, t, hours
t_{Sprf}	=	start of pseudo radial flow, t, hours
t_s	=	time required for stabilization, hours
T	=	reservoir temperature, °R
T_{sc}	=	standard condition temperature, °R
u	=	dummy variable
V	=	volume, bbl
V_f	=	fraction of bulk volume occupied by fractures
V_m	=	fraction of bulk volume occupied by matrix
V_w	=	V_{wb} = wellbore volume, bbl
w	=	width of channel reservoir, feet

(*Continued*)

Table 1.12 Cont.

w_f	=	fracture width, feet
wk_f	=	fracture conductivity, md-feet
w_s	=	width of damaged zone around fracture face,
WBS	=	wellbore storage
z	=	gas-law deviation factor, dimensionless
$\bar{z}$	=	gas-law deviation factor at average reservoir pressure, dimensionless
Δp	=	pressure change since start of transient test, psi
$(\Delta p)_{MP}$	=	pressure change at match point
Δp_D	=	dimensionless pressure change
Δp_p	=	pseudopressure change since start of test, psia²/cp
Δp_s	=	additional pressure drop due to skin, psi
$\Delta p_{t=0}$	=	pressure drop at time zero, psi
Δp_{1hr}	=	pressure change from start of test to one hour elapsed time, psi
Δt	=	time elapsed since start of test, hours
Δt_a	=	$\overline{\mu c_t}\Delta t_{ap}$, normalized or adjusted pseudo time, hours
Δt_{ap}	=	$\int_0^{\Delta t} \frac{dt}{\mu(p)c_t(p)}$, pseudo time, hr-psia/cp
Δt_{Be}	=	bilinear equivalent time, hours
Δt_e	=	radial equivalent time, hours
Δt_{Le}	=	linear equivalent time, hours
Δt_{max}	=	maximum shut-in time in pressure buildup test, hours
ΔV	=	change in volume, bbl
η	=	0.0002637 $k/\phi\mu c_t$, hydraulic diffusivity, feet²/hr
η_{fD}	=	hydraulic diffusivity, dimensionless
λ	=	interporosity flow coefficient
λ_t	=	$\frac{k_0}{\mu_0}+\frac{k_w}{\mu_w}+\frac{k_g}{\mu_g}$, total mobility, md/cp
α	=	exponent in deliverability equation
α	=	parameter characteristic of system geometry in dual-porosity system

(*Continued*)

Table 1.12 Cont.

β	=	turbulence factor
β'	=	transition parameter
γ	=	Euler's constant, = 1.781, dimensionless
γ_g	=	gas gravity (air = 1.0)
γ_m	=	matrix density
ω	=	storativity ratio in dual porosity reservoir
μ	=	viscosity, cp
μ_i	=	viscosity evaluated at p_i, cp
μ_g	=	gas viscosity, cp
μ_o	=	oil viscosity, cp
μ_w	=	water viscosity, cp
$\bar{\mu}_g$	=	gas viscosity evaluated at average pressure, cp
μ_{gwf}	=	gas viscosity evaluated at p_{wf}, cp
$\bar{\mu}$	=	viscosity evaluated at $\bar{p}$, cp
ρ	=	density, lbm/feet^3 or g/cm^3
ρ_{wb}	=	density of liquid in wellbore, lbm/feet^3
ϕ_f	=	fraction of fracture volume occupied by pore space, ≈ 1
ϕ_m	=	fraction of matrix volume occupied by pore space
$(\phi V)_f$	=	fraction of bulk volume occupied by pore space in fractures
$(\phi V c_t)_f$	=	fracture "storativity" for dual porosity reservoir
$(\phi V c_t)_{f+m}$	=	total "storativity" for dual porosity reservoir
$(\phi V)_m$	=	fraction of bulk volume occupied by pore space in matrix
ϕ	=	porosity, dimensionless
Σs	=	sum of damage skin, turbulence, and other pseudo skin factors

offset wells). In pseudo steady-state, reservoir pressure drops at the same rate with time at all points in the reservoir, including at the reservoir boundaries. Ideally, true steady-state flow can occur in the drainage area of a well, but only if pressure at the drainage boundaries of the well can be maintained constant while the well is producing at constant rate. While unlikely, steady-state flow is conceivable for wells with edge water drive or in repeated flood patterns in a reservoir. The solution to the radial diffusivity equation is based on a constant-pressure outer boundary condition, instead of a no-flow outer boundary condition. The steady-state solution, applicable after boundary effects have been felt, is:

$$p_i - p_{wf} = 141.2(qB\mu/kh)[\ln(r_e/r_w) + s]$$

1.22.3 Linear Flow

Linear flow occurs in some reservoirs (i) with long, highly conductive vertical fractures, (ii) in relatively long, relatively narrow reservoirs – channels, such as ancient stream beds, and (iii) in near horizontal wells during certain times. For unsteady-state linear flow in an unbounded (infinite-acting) reservoir:

$$p_{wf} = p_i - 16.26(qB\mu/kA)(kt/\varphi\mu c_t)^{1/2} = 70.6(qB\mu/kh)s_f$$

1.22.4 Spherical Flow

Spherical flow occurs in wells with limited perforated intervals and into wireline formation test tools. The solution to the spherical/cylindrical, one-dimensional form of the diffusivity equation, subject to the initial condition that pressure is uniform before production and the boundary conditions of constant flow rate and an infinitely large drainage area, is:

$$p_{wf} = p_i - (70.6qB\mu)/(k_s r_s) + [2456(\varphi\mu c)^{1/2}qB\mu/ k_{sp}^{3/2} r_{sp}]1/t^{1/2} - [(70.6qB\mu)/9k_{sp} r_{sp})]s$$

$$k_{sp} = (k_r k_z^{1/2})^{2/3}$$

and r_{sp} = the radius of the sphere into which flow converges.

1.23 Flow Through Porous Media

When oil is produced from a well, the oil first flows through the formation or the sandstone to the well. The formation is a porous matrix that allows fluid to passing through. For single fluid permeating through a vastly unbounded porous media, the governing equation remains the one first conceived by Darcy in 1856.

In an extension of Darcy's law to multiphase flows, the equation remains the same for each phase but allows the fluid properties as well as the permeability to differ. That is:

$$Vi = (k_i/\mu_i)(Vp_i - \rho_i g)$$

The subscript *i* denotes for the ith fluid phase.

For multiphase flows, the flow of one phase can affect the motion of other phases. It may be expected that the interactions are linear when the inertia is negligible. For two-phase flows through porous media, the phase interactions may be added:

$$-(\nabla p_i - \rho_i g) = \mu_i (v_i/k_i - v_j/k_{ij})$$

k_{ij} is the phase interaction coefficient.

1.24 Flow Velocity

Flow velocity, V, is calculated from:

$$V = Q/A$$

Q is the flow rate and A is the cross-sectional area of the pipe.

When the flow rate is in gallons per minute and the cross-sectional area is in square inches:

$$\text{V in feet per minute} = (19.5 \times Q)/A$$

When the flow rate is in barrels per minute and the cross-sectional area is in square inches:

$$\text{V in feet per minute} = (808.5 \times Q)/A$$

Example calculation

Calculate the fluid velocity inside the pipe as well as in the annulus with the dimensions as follows for a flow rate of 350 gpm (4.762 bpm) if the pipe inside diameter is 3 inches, the pipe outside diameter is 4.5 inches, and the hole diameter is 8.5 inches.

Solution:

The velocity inside the pipe using flow rate in gpm is:

$$V_p = (19.25 \times 200)/[\pi/4 \times (8.5^2 - 4.5^2] = 94.3 \text{ fpm}$$

The velocity inside pipe using flow rate in bpm is:

$$V_p = (808.5 \times 4.762)/[\pi/4 \times 3^2] = 544.7 \text{ fpm}$$

1.25 Fluid Saturation

Fluid saturation is the petrophysical property that describes the amount of each fluid type in the pore space. It is defined as the fraction of the pore space occupied by a fluid phase. In general,

$$\text{Fluid Saturation} = (\text{Fluid volume})/(\text{effective rock pore volume})$$

All saturation values are based on pore volume and not on the gross reservoir volume. The saturation of each individual phase ranges between zero to 100 percent. By definition, the sum of the saturations is 100%, therefore:

$$Sg + So + Sw = 1.0$$

Sg = volume of gas/pore volume, So = volume of oil/pore volume, Sw = volume of water/pore volume.

1.26 Formation Volume Factor – Gas

The gas formation volume factor is used to relate the volume of gas, as measured at reservoir conditions, to the volume of gas as measured at standard conditions (60°F, 14.7 psia). It is the ratio of volume of 1 mol of gas at a given pressure and temperature to the volume of 1 mole of gas at standard conditions (p_s and *Ts*). Using the real gas law and if the Z factor at standard conditions is 1, the equation for formation volume factor (B_g) is:

$$B_g = V_R/V_S = (nZRT/P)(P_s/nZ_sRT_s) = P_sZT/T_sP$$

When P_s is 1 atmosphere (14.6959 psia or 101.325 kPa) and T_s is 60°F (519.67°R or 288.71°K), this equation can be written in other forms:

$$B_g = 0.0282793(ZT/p)\text{rcf/scf}$$

$$B_g = 0.00503676(ZT/p)\text{RB/scf}$$

$$B_g = 0.350958(ZT/p)\text{Rm}^3/\text{Sm}^3$$

where rcf/scf = reservoir cubic feet per standard cubic feet, RB = reservoir barrels, and Rm^3/Sm^3 = reservoir cubic meters per standard cubic meters.

1.27 Formation Volume Factor – Oil

The formation volume factor for oil (B_o) is the volume in barrels that one stock tank barrel occupies in the formation at reservoir temperature and with the solution gas that is held in the oil at reservoir pressure. Due to the dramatically different conditions prevailing at the reservoir when compared to the conditions at the surface, it is not expected that 1 barrel of fluid at reservoir conditions could contain the same amount of matter as 1 barrel of fluid at surface conditions. Thus:

$$B_o = (Vo)_{p,T}/V_o)_{SC}$$

In this equation, B_o is the oil formation volume factor in bbl/STB, $(V_o)_{p,T}$ is the volume of oil (bbls) under reservoir pressure (p) and temperature (T), and $(V_o)_{SC}$ is the volume of oil (bbls) measure under standard conditions (STB).

Values typically range from approximately 1.0 bbl/STB for crude oil systems containing little or no solution gas to nearly 3.0 bbl/STB for highly volatile oils. For saturated systems, gas is liberated as pressure is reduced below the bubblepoint which results in a shrinkage in oil volume. Generally:

Dead oil (no dissolved gas): B_o = approximately 1.0

Gassy (deep) oil: B_o = approximately 1.4

Typical (shallow) oil: B_o = approximately 1.2

1.28 Friction

1.28.1 Coefficient of Friction

The coefficient of friction is the ratio of the frictional force to the normal force acting at the point of contact. It is given as

$$\mu = F_f/F_n$$

where F_f = friction force, lbf, and F_n = normal force, lbf.

The coefficient of friction is a scalar dimensionless value that depends on the surface but is independent of the surface area.

1.28.2 Types of Friction

Static friction:

$$\mu_s = F_{sf}/F_n$$

Kinetic friction:

$$\mu_k = F_{kf}/F_n$$

Rolling fraction:

$$\mu_r = F_{rf}/F_n$$

1.28.3 Friction and Rotational Speed

The following empirical equation provides a good representation and coupling of the friction effects and drill string rotating speed as well as tripping speed:

$$\mu_v = \mu_s \times e^{-kVrs}$$

The resultant velocity, V_{rs}, of a contact point on the drill string is the vector sum of two components: circumferential velocity V_C (caused by rotation) and axial velocity V_{ts} (affected by drilling rate or tripping speed).

The friction factor, which has the dependency on the side force, kinematics, temperature, and geometrical parameters of the contacting surfaces, is given by:

$$M_v = [\mu_s/1 + (\mu_s \sigma_n/k\Delta t)V_{rs}]$$

where σ_n is the normal stress at the contact, Δt is the average contact temperature, V_{ts} is the trip speed, V_{rs} is the resultant speed.

1.29 Gas Deviation Factor

A natural gas mixture under reservoir conditions is non-ideal and the behavior can be approximated by the real gas law, a general equation of state for gases:

$$pV = ZnRT$$

In this equation, p is pressure in psi, V is the gas volume in cubic feet, n is the number of moles of the gas, T is absolute temperature in R (degrees Rankine), R is the universal gas constant and equals to 10.73 psi ft3/lb-mol-R, and Z is the gas deviation factor or *Z-factor*, which may also be called the *super-compressibility factor* and is rhe ratio of the real volume (the volume actually occupied by a gas at a given p and T) to the ideal volume (volume it would occupy had it behaved as an ideal gas). The Z *factor* can be determined in a PVT laboratory. In common practice it is calculated from published charts.

1.30 Gas Solubility

The gas solubility is the number of standard cubic feet of gas that will dissolve in one stock-tank barrel of crude oil at certain pressure and temperature. The solubility of natural gas in a crude oil is a strong function of the pressure, temperature, API gravity, and gas gravity. For a dilute solution, the partial pressure exerted by a dissolved liquid (a solute) a in a liquid solvent is given by:

$$p_a = Hx_a$$

H is Henry's law constant for the system and xa is the mole fraction of solute. A different value of H is applicable to each gas–liquid system.

The following empirical correlations for estimating the gas solubility are: (1) Standing's correlation, (2) Vasquez-Biggs correlation, (3) Glaso's correlation, (4) Marhoun's correlation, and (5) Petrosky-Farshad correlation.

1.31 Gas-Oil Ratio

The produced gas-oil ratio (GOR) at any particular time is the ratio of the standard cubic feet of *total* gas being produced at any time to the stock-tank barrels of oil being produced at that same instant. Hence, the name *instantaneous gas-oil ratio* is described the GOR mathematically by the following expression:

$$GOR = R_s + (k_{rg}/k_{ro})[(\mu_o B_o/\mu_g B_g]$$

GOR is the instantaneous gas-oil ratio, scf/STB R is the gas solubility, scf/STB, k_{rg} is the relative permeability to gas, k_{ro} is the relative permeability to oil, B_o is the oil formation volume factor, bbl/STB, B_g is the gas formation volume factor, bbl/scf, μ_o is the oil viscosity, cp, and μ_g is the gas viscosity, cp

1.32 Geothermal Gradient

The geothermal gradient is the rate of increasing temperature with respect to increasing depth in the interior of the Earth. Away from tectonic plate boundaries, the gradient is approximately 25°C per kilometer of depth (1°F per 70 feet of depth) near the surface in most of the world. However, the geothermal gradient varies with location and is typically measured by determining the bottom open-hole temperature after borehole drilling. Although the geothermal gradient varies from place to place, it is generally on the order of 25 to 30°C/km (15°F/1000 ft or 120°C/1000 feet, i.e. 0.015°C per foot of depth or 0.012°C per foot of depth).

In the geosciences, the measurement of temperature (T) is associated with heat flow (Q):

$$Q = K\Delta T/\Delta Z$$

K is the thermal conductivity of the rock.

1.33 Hole Capacity

Hole diameter (in)	Capacity (bbl/ft)	Capacity (ft/bbl)
3	0.0087	114.387
3-1/8	0.0095	105.419
3-1/4	0.0103	97.466
3-3/8	0.0111	90.380
3-1/2	0.0119	84.040
3-5/8	0.0128	78.344
3-3/4	0.0137	73.208
3-7/8	0.0146	68.561
4	0.0155	64.343
4-1/8	0.0165	60.502
4-1/4	0.0175	56.996

(Continued)

Hole diameter (in)	Capacity (bbl/ft)	Capacity (ft/bbl)
4-3/8	0.0186	53.785
4-1/2	0.0197	50.839
4-5/8	0.0208	48.128
4-3/4	0.0219	45.628
4-7/8	0.0231	43.318
5	0.0243	41.179
5-1/8	0.0255	39.195
5-1/4	0.0268	37.351
5-3/8	0.0281	35.634
5-1/2	0.0294	34.033
5-5/8	0.0307	32.537
5-3/4	0.0321	31.138
5-7/8	0.0335	29.827
6	0.0350	28.597
6-1/8	0.0364	27.442
6-1/4	0.0379	26.355
6-3/8	0.0395	25.331
6-1/2	0.0410	24.367
6-5/8	0.0426	23.456
6-3/4	0.0443	22.595
6-7/8	0.0459	21.781
7	0.0476	21.010
7-1/8	0.0493	20.279
7-1/4	0.0511	19.586
7-3/8	0.0528	18.928
7-1/2	0.0546	18.302
7-5/8	0.0565	17.707
7-3/4	0.0583	17.140
7-7/8	0.0602	16.600
8	0.0622	16.086
8-1/8	0.0641	15.595
8-1/4	0.0661	15.126
8-3/8	0.0681	14.677

(Continued)

Hole diameter (in)	Capacity (bbl/ft)	Capacity (ft/bbl)
8-1/2	0.0702	14.249
8-5/8	0.0723	13.839
8-3/4	0.0744	13.446
8-7/8	0.0765	13.070
9	0.0787	12.710
9-1/8	0.0809	12.364
9-1/4	0.0831	12.032
9-3/8	0.0854	11.713
9-1/2	0.0877	11.407
9-5/8	0.0900	11.113
9-3/4	0.0923	10.830
9-7/8	0.0947	10.557
10	0.0971	10.295
10-1/8	0.0996	10.042
10-1/4	0.1021	9.799
10-3/8	0.1046	9.564
10-1/2	0.1071	9.338
10-5/8	0.1097	9.119
10-3/4	0.1123	8.908
10-7/8	0.1149	8.705
11	0.1175	8.508
11-1/8	0.1202	8.318
11-1/4	0.1229	8.134
11-3/8	0.1257	7.956
11-1/2	0.1285	7.784
11-5/8	0.1313	7.618
11-3/4	0.1341	7.457
11-7/8	0.1370	7.301
12	0.1399	7.149
12-1/8	0.1428	7.003
12-1/4	0.1458	6.860
12-3/8	0.1488	6.722
12-1/2	0.1518	6.589

(Continued)

Hole diameter (in)	Capacity (bbl/ft)	Capacity (ft/bbl)
12-5/8	0.1548	6.459
12-3/4	0.1579	6.333
12-7/8	0.1610	6.210
13	0.1642	6.092
13-1/8	0.1673	5.976
13-1/4	0.1705	5.864
13-3/8	0.1738	5.755
13-1/2	0.1770	5.649
13-5/8	0.1803	5.546
13-3/4	0.1836	5.445
13-7/8	0.1870	5.348
14	0.1904	5.252
14-1/8	0.1938	5.160
14-1/4	0.1972	5.070
14-3/8	0.2007	4.982
14-1/2	0.2042	4.896
14-5/8	0.2078	4.813
14-3/4	0.2113	4.732
14-7/8	0.2149	4.653
15	0.2186	4.575
15-1/8	0.2222	4.500
15-1/4	0.2259	4.427
15-3/8	0.2296	4.355
15-1/2	0.2334	4.285
15-5/8	0.2371	4.217
15-3/4	0.2410	4.150
15-7/8	0.2448	4.085
16	0.2487	4.021
16-1/8	0.2526	3.959
16-1/4	0.2565	3.899
16-3/8	0.2605	3.839
16-1/2	0.2645	3.781

(*Continued*)

Hole diameter (in)	Capacity (bbl/ft)	Capacity (ft/bbl)
16-5/8	0.2685	3.725
16-3/4	0.2725	3.669
16-7/8	0.2766	3.615
17	0.2807	3.562
17-1/4	0.2890	3.460
17-1/2	0.2975	3.362
17-3/4	0.3060	3.268
18	0.3147	3.177
18-1/4	0.3235	3.091
18-1/2	0.3324	3.008
18-3/4	0.3415	2.928
19	0.3507	2.852
19-1/4	0.3599	2.778
19-1/2	0.3694	2.707
19-3/4	0.3789	2.639
20	0.3885	2.574
20-1/4	0.3983	2.511
20-1/2	0.4082	2.450
20-3/4	0.4182	2.391
21	0.4284	2.334
21-1/4	0.4386	2.280
21-1/2	0.4490	2.227
21-3/4	0.4595	2.176
22	0.4701	2.127
22-1/4	0.4809	2.080
22-1/2	0.4918	2.034
22-3/4	0.5027	1.989
23	0.5138	1.946
23-1/4	0.5251	1.904
23-1/2	0.5364	1.864
23-3/4	0.5479	1.825
24	0.5595	1.787

(Continued)

Hole diameter (in)	Capacity (bbl/ft)	Capacity (ft/bbl)
24-1/4	0.5712	1.751
24-1/2	0.5831	1.715
24-3/4	0.5950	1.681
25	0.6071	1.647
26	0.6566	1.523
27	0.7081	1.412
28	0.7615	1.313
29	0.8169	1.224
30	0.8742	1.144
31	0.9335	1.071
32	0.9947	1.005

1.34 Horsepower

$$HP = (\text{force, pound-foot} \times \text{distance, feet})/55 \times \text{time, seconds}$$

Also:

$$HP = (\text{force, pound-foot} \times \text{velocity, feet/minutes})/33{,}000$$

Hydraulic horsepower is determined by the equation:

$$HHP = (\text{flow rate, gallons per minute} \times \text{pressure, psi})/1714$$

Rotating horsepower is determined by the equation:

$$HP = (\text{torque, ft-lbf} \times \text{speed, rpm})/5252$$

One horsepower = 550 foot-pounds second^{-1} = 33,000 ft-lbf/minute, and 33,000 ft-lbf/6.2832 = 5252. Other conversion factors are: 1 horsepopwer = 0.0007457 megawatts = 0.7457 kilowatts = 745.7 watts.

Example calculation

The torque on a motor shaft follows a sinusoidal pattern with a maximum amplitude of 8,000 ft-lbf. The torque will always be positive through the cycle. Shaft speed is 200 rpm. Calculate the motor horsepower.

Solution:
Horsepower is given as

$$HP = 2\pi NT/3300$$

The torque is calculated using the sinusoidal pattern:

$$T = \int_0^{\pi} 8000 \sin x dx,$$

$$T = 2\left[8000 \cos x\right]_0^{\pi}.$$

Substituting the limits:

$$T = 16000[\cos \pi - \cos \theta]^{\circ}\pi = 32000 \text{ ft-lbf}$$

$$HP = (2\pi \times 200 \times 32000)/33000 - 1219 \text{ hp}$$

1.35 Hydrostatic Pressure

The hydrostatic pressure is the pressure created by a column of fluid due to its density and vertical height. This type of pressure always exists and may be calculated whether the fluid is static or flowing. It can be calculated using the following relationship:

$$_{H}p = MW \times 0.0519 \times TVD$$

H_p is the hydrostatic pressure in psi, MW is the mud density in lbs per gallon, and TVD is the True Vertical Depth in feet, i.e. the vertical distance from the depth reference level to a point on the borehole course.

1.36 Isothermal Compressibility of Oil

The isothermal compressibility of is the change in volume of a system as the pressure changes while temperature remains constant. For undersaturated oil, the isothermal compressibility is defined by the following equation:

$$c_o = -1/V(\delta V/\delta p)_T = -1/B_o(\delta B_o/\delta p)_T$$

This equation reflects the change in volume with change in pressure under constant temperature conditions.

Below the bubble point pressure, isothermal compressibility of oil is defined from oil and gas properties to account for gas coming out of solution. Thus, for saturated oil compressibility is given by:

$$c_o = -1/B_o(\delta B_o/\delta p)_T - 1/B_g(\delta R_s/\delta p)_T$$

Above the bubble point pressure, oil volume changes as a function of isothermal compressibility only.

The oil formation volume factor (FVF) for undersaturated crude oil is determined as a function of bubblepoint oil formation volume factor, isothermal compressibility, and pressure above bubblepoint from:

$$B_o = B_{ob} e^{[co(pb\text{-}p)]}$$

Symbols

c_o = oil isothermal compressibility, Lt^2/m, psi^{-1}
B_o = oil FVF, bbl/STB
B_g = gas FVF, ft^3/scf
T = temperature, T, °F
R_s = solution GOR, scf/STB
p = pressure, m/Lt^2, psia
B_{ob} = oil formation volume at bubblepoint pressure, bbl/STB
p_b = bubblepoint pressure, m/Lt^2, psia

1.37 Marx-Langenheim Model

The Marx-Langenheim model is a series of mathematical equations for calculating heat transfer in a hot water or steam flood. Many of currently available simplified methods are based on the reservoir heating model of Marx and Langenheim which considers the injection of hot fluid into a well at constant rate and temperature (Figure 1.2). The operation element consists of a radial flow system, concentric about the point of injection. They assumed the temperature of the heated zone to be uniform at the downhole temperature of the injected fluid (*Ts*) and the reservoir temperature outside the heated zone to be at the initial and reference temperature (*TR*).

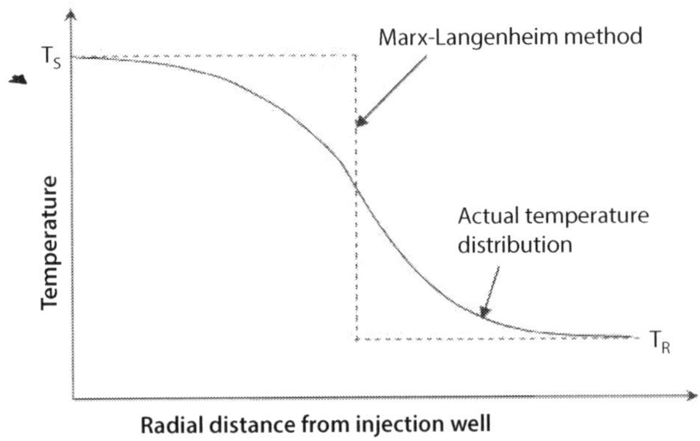

Figure 1.2 Schematic of the Marx-Langenheim Temperature Profile.

1.38 Material Balance

Material balance (*mass balance*) is an application of conservation of mass an expression for conservation of mass governed by the observation that the amount of mass leaving a control volume is equal to the amount of mass entering the volume minus the amount of mass accumulated in the volume.

$$\text{M (original)} = \text{M (remaining)} + \text{M (removed)}$$

Furthermore, the pressures measured over time can be used to estimate the volume of hydrocarbons in place.

The *material balance method* for a crude oil and natural gas field uses an equation that relates the volume of oil, water and gas that has been produced from a reservoir, and the change in reservoir pressure, to calculate the remaining oil. It assumes that as fluids from the reservoir are produced, there will be a change in the reservoir pressure that depends on the remaining volume of oil and gas. The method requires extensive pressure-volume-temperature analysis and an accurate pressure history of the field. It requires some production to occur (typically 5% to 10% v/v of ultimate recovery), unless reliable pressure history can be used from a field with similar rock and fluid characteristics.

Assumptions that can be made about material balance equations are: (1) masses and volumes of components are additive, (2) material is neither generated nor lost from the system, (3) for a three-component mixture of oil (o), water (w) and solids (s), where:

$$MW = D_s V_s + D_o V_o + D_w V_w$$

$$V_s + V_o + V_w = 100\%$$

V is the volume percent, D is the specific gravity, and MW is the mixture weight). (4) there is uniform distribution of pressures and saturations, (5) all areas are can be swept or reached by fluids, (6) the reservoir properties are homogeneous properties, and (7) the reservoir is a tank.

Commonly used variables in material balance equations relate to barite, hematite and (light) crude oil (Table 1.13), although in the case of other

Table 1.13 Commonly Used Variables in Material Balance Equations.

Barite
1. Weight of a barrel of barite (barium sulfate, $BaSO_4$), specific gravity: 4.2 g/cc
42 gal/bbl × 8.33 lb/gal × 4.2 = 1470 lb/bbl
2. Weight of a gallon of barite: 8.33 lb/gal × 4.2 = 34.9 lb/gal
Hematite
1. Weight of a barrel of hematite (ferric oxide, Fe_2O_3), specific gravity: 5.0 g/cc
42 gal/bbl × 8.33 lb/gal × 5.0 = 1749 lb/bbl
2. Weight of a gallon of hematite: 8.33 lb/gal × 5.0 = 41.65 lb/gal
Crude oil
1. Light crude oil – 41° API Gravity, specific gravity: 0.82 g/cc
Weight of a gallon of light crude oil: 8.33 lb/gal × 0.82 = 6.8 lb/gal
2. Medium crude oil – 22.3° API gravity, specific gravity: 0.92 g/cc
Weight of a gallon of medium crude oil: 8.33 lb/gal × 0.92 = 7.7 lb/gal
3. Heavy crude oil – 18° API gravity, specific gravity: 0.95 g/cc
Weight of a gallon of heavy crude oil: 8.33 lb/gal × 0.95 = 7.9 lb/gal
4. Extra heavy oil – 8° API gravity, specific gravity: 1.01 g/cc
Weight of a gallon of extra heavy crude oil: 8.33 lb/gal × 1.01 = 8.4 lb/gal

crude oil feedstocks, adjustments will to be made for the increasing use of heavier feedstocks in recovery operations:

1.39 Modulus of Elasticity

The modulus of elasticity is given by the relationship:

$$E = \sigma/\varepsilon = (F/A)/\Delta L/L \text{ psi}$$

Σ is the unit stress, ε is the unit strain per inch per inch, F is the axial force, lbf, A is the cross-sectional area in square inches, ΔL is the total strain or elongation in inches, and L is the original length in inches.

1.40 Oil and Gas Originally in Place

The *oil originally in place* (OOIP) is the quantity of petroleum existing in a reservoir before oil recovery operations begin. OOIP is usually reported in the units of stock-tank barrels. A *stock-tank barrel* (stb) refers to a barrel of oil at surface standard conditions (42 U.S. gallons, 5.615 cu. ft.).

For oil reservoirs, the original oil in place (OOIP) volumetric calculation based on the metric system is:

$$\text{OOIP } (m^3) = \text{Rock Volume} \times \text{Ø} \times (1 - S_w) \times 1/B_o$$

Where: the rock volume (m^3) is $10^4 \times A \times h$ – A is the drainage area, hectares (1 ha = 10^4 m^2), h is the net pay thickness in meters – Ø is the porosity, which is the fraction of rock volume available to store fluids, S_w is the volume fraction of porosity that is filled with interstitial water, B_o is the formation volume factor (m^3/m^3) which is a dimensionless factor for the change in oil volume between reservoir conditions and standard conditions at surface, $1/B_o$ is the shrinkage (stock tank m_3/reservoir m^3) is the volume change that the oil undergoes when brought to the surface facilities and is due to solution gas evolving out of the oil.

For oil reservoirs the original oil in place (OOIP) volumetric calculation based on the imperial system is:

$$\text{OOIP (STB)} = \text{rock volume} \times 7{,}758 \times \text{Ø} \times (1 - S_w) \times 1/B_o$$

Where: the rock volume (acre feet) is A × h – A is the drainage area in acres, h is the net pay thickness in feet, 7,758 is the factor that converts acre-feet to stock tank barrels, Ø is the porosity, which is the fraction of rock volume available to store fluids, S_w is the volume fraction of porosity that is filled with interstitial water, B_o is the formation volume factor (reservoir bbl/STB), $1/B_o$ is the shrinkage (stb/reservoir bbl).

To convert the estimates to stock tank barrels, the equation is:

$$STOIIP = n = v\varphi(1 - S_{wc})/B_{oi}$$

In this equation, B_{oi} is the oil formation volume factor, under initial conditions, V is the net bulk volume of the reservoir rock, φ is the porosity that is presented as a fractional value, S_{wc} is the irreducible or connate water saturation (fraction), B_{oi} is the initial oil formation volume factor (rb/stb), and V_b denotes the bulk volume of the reservoir and is calculated from estimates of the often complex lateral dimensions and thicknesses of the reservoir. The STOIIP value have the units: reservoir volume/stock tank volume, usually, reservoir barrels/stock tank barrel (rb/stb). Thus a volume of B_{oi} rb of oil will produce one stb of oil at the surface together with the volume of gas which was originally dissolved in the oil in the reservoir.

To calculate recoverable oil volume of oil, the estimate of the original oil in place must be multiplied by the recovery factor. Fluid properties such as formation volume factor, viscosity, density, and solution gas/oil ratio all influence the recovery factor. In addition, it is also a function of the reservoir drive mechanism and the interaction between reservoir rock and the fluids in the reservoir. Some industry standard oil recovery factor ranges for various natural drive mechanisms are: (1) solution gas drive: 2 to 30% v/v, (2) gas cap drive: 30 to 60% v/v, (3) water drive: 2 to 50% v/v, and gravity drive: up to 60% v/v.

For gas reservoirs the original gas-in-place (OGIP) volumetric calculation using the metric system is:

$$OGIP\ (10^3\ m^3) = \text{rock volume} \times Ø \times (1 - s_w) \times (T_s \times Pi)/(P_s \times T_f \times Z_i)$$

1.41 Oil Recovery Factor

Simply, the oil recovery factor (ORF) is the estimate of the recoverable oil (ERO) divided by the estimate of the in-place oil (EIPO):

$$ORF = ERO/EIPO$$

On the other hand, the equation of the stock tank oil initially in place (STOIIP) can be converted into an equation for calculating the ultimate oil recovery by multiplying by the recovery factor (RF), which is a number between zero and unity representing the fraction of recoverable oil. Thus:

$$\text{Ultimate recovery (UR)} = [(V\varphi\ (1 - S)/B_{oi}] \times RF$$

1.42 Permeability

Permeability (commonly symbolized as κ, or *k*) is a measure of the ease with which fluid is able to flow through a porous medium (in the current context, a rock or unconsolidated material) to transmit fluids. It is of great importance in determining the flow characteristics of hydrocarbons in oil and gas reservoirs (Table 1.14). It is typically measured through calculation of Darcy's law. The intrinsic permeability of any porous material is:

$$k_I = Cd^2$$

κ_I is the intrinsic permeability, C is a dimensionless constant that is related to the configuration of the flow-paths, d is the average, or effective pore diameter [L]. A common unit for permeability is the Darcy (D), or more commonly the milliDarcy (mD). Typically, reservoir rock permeability lies in the range of a few milliDarcys to several Darcys, but exceptions are not uncommon. Productive limestone matrix permeability may be as low as a fraction of a milliDarcy while fractures, vugs, and caverns may have extremely high permeability.

Variation of permeability within a reservoir is common, with vertical permeability (k_v) typically being lower than horizontal permeability (k_h) due to stratification and layering that occurs during reservoir deposition. Vertical-to-horizontal permeability ratios (k_v/k_h) typically lie in the range of 0.1 to 0.001. A situation in which horizontal permeability in one direction varies from that in another direction is known as permeability anisotropy.

1.43 Poisson's Ratio

Poisson's ratio is given by the equation:

$$\nu = \varepsilon_{lat}/\varepsilon_{long}$$

Table 1.14 Permeability of Different Systems.

<table>
<tr><th>Permeability</th><th colspan="4">Pervious</th><th colspan="4">Semi-pervious</th><th colspan="5">Impervious</th></tr>
<tr><td>Unconsolidated Sand and Gravel</td><td colspan="2">Well Sorted Gravel</td><td colspan="3">Well Sorted Sand; Sand and Gravel</td><td colspan="4">Very Fine Sand, Silt, Loess, Loam</td><td colspan="4"></td></tr>
<tr><td>Unconsolidated Clay</td><td colspan="4"></td><td colspan="2">Peat</td><td colspan="3">Layered Clay</td><td colspan="4">Fat/Unweathered Clay</td></tr>
<tr><td>Consolidated Rocks</td><td colspan="4">Highly Fractured Rocks</td><td colspan="3">Reservoir Rocks</td><td colspan="2">Fresh Sandstone</td><td colspan="2">Limestone; Dolomite</td><td colspan="2">Granite</td></tr>
<tr><td>κ (cm^2)</td><td>0.001</td><td>0.0001</td><td>10^{-5}</td><td>10^{-6}</td><td>10^{-7}</td><td>10^{-8}</td><td>10^{-9}</td><td>10^{-10}</td><td>10^{-11}</td><td>10^{-12}</td><td>10^{-13}</td><td>10^{-14}</td><td>10^{-15}</td></tr>
<tr><td>κ (millidarcy)</td><td>10^{+8}</td><td>10^{+7}</td><td>10^{+6}</td><td>10^{+5}</td><td>10,000</td><td>1,000</td><td>100</td><td>10</td><td>1</td><td>0.1</td><td>0.01</td><td>0.001</td><td>0.0001</td></tr>
</table>

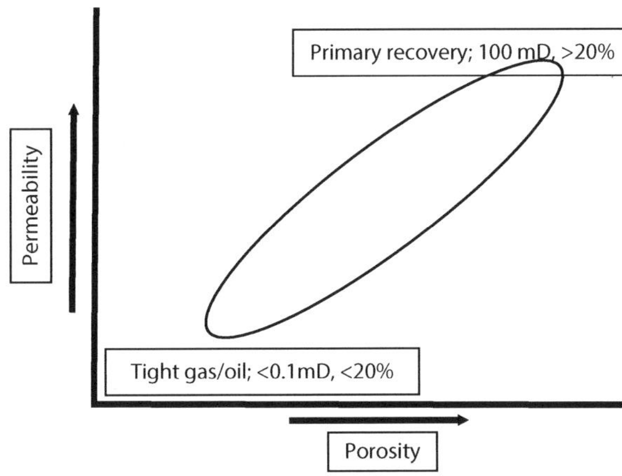

Figure 3 General Trends in the Relationship Between Porosity and Permeability.

Table 1.15 The modulus of elasticity, the shear modulus, and Poisson's ratio for common metals at room temperature.

Metal alloy	Modulus of elasticity		Shear modulus		Poisson's ratio
	Psi × 10⁶	MPa × 10⁶	Psi × 10⁶	MPa × 10⁶	
Aluminum	10	6.9	3.8	2.6	0.33
Copper	16	11	6.7	4.6	0.35
Steel	30	20.7	12	8.3	0.27
Titanium	155	10.7	6.5	4.5	0.36
Tungsten	59	40.7	23.2	16	0.28

ε_{lat} is the lateral strain in inches and ε_{long} is the longitudinal or axial strain in inches. The modulus of elasticity, the shear modulus, and Poisson's ratio for most metals, the Poisson ration varies from 0.25 to 0.33 (Table 1.15).

The modulus of elasticity, E, and the shear modulus, G, are related to the Poisson ratio by:

$$E = 2G(1 + \nu).$$

1.44 Porosity

Porosity is a measure of the spaces between grains of sediment in sedimentary rock. A porous system is a system that allows storage (porosity) and enables transmission (permeability). The porosity of the reservoir rock is a measure of the pore space available for the storage of fluids in rock. Thus:

$$\text{Porosity, f} = Vp/Vb = (Vb - Vm)/Vp$$

The porosity is expressed as a fraction of per cent, Vb = Vp + Vm, Vb is the bulk volume of reservoir rock (liters^3), Vp is the pore volume, (liters^3), and Vm is the matrix volume, (liters^3).

The *total porosity* is the amount of void space in a formation rock, usually expressed as a per cent of the voids and/or spaces per bulk volume. Total porosity is the total void space in the rock whether or not it contributes to fluid storage, and thus can include isolated pores and the spaces occupied by clay bound water. Porosity is expressed as a fraction between 0 and 1 or as a percentage between 0 and 100%.

$$\text{Total porosity} = \text{Effective porosity} + \text{Ineffective porosity}$$

The total porosity is often referred to as the *absolute porosity*. Thus, the *absolute porosity* is the ratio of the total pore space the rock to that of the bulk volume. whether or not that space is accessible to fluid penetration. A sediment may have considerable absolute porosity and yet have no conductivity to fluid for lack of pore.

$$\Phi_a = \text{total pore volume/bulk volume}$$

or

$$\Phi_a = (\text{bulk volume} - \text{grain volume})/\text{total bulk volume}$$

or

$$\Phi_a = \frac{\text{vol of interconnected pores} + \text{vol of dead end pores} + \text{vol of isolated pores}}{\text{total or bulk volume}}$$

Effective porosity refers to the amount of inter-connected pore spaces, i.e., the space available to fluid penetration, in relation to the bulk volume. It includes cul-de-sac or dead end porosity. Therefore, *ineffective porosity* is the amount of pore spaces that are not interconnected i.e. the isolated pore spaces, in relation to the bulk volume.

Effective porosity is given by:

$$\Phi_e = \text{interconnected pore volume/bulk volume}$$

or

$$\Phi_e = (\text{volume of interconnected pores} + \text{cul-de-sacs})/\text{total or bulk volume}$$

Ineffective porosity is given by:

$$\Phi_{ineff} = (\text{volume of completely disconnected pores})/\text{total volume or bulk volume}$$

Other definitions include (1) primary porosity or original porosity, which developed at time of deposition, and (2) the secondary porosity, which developed as a result of geologic processes occurring after deposition.

1.45 Pressure Differentials

When two columns of fluid exist, one in the annulus and one inside the tubing, they are in hydraulic connection, but substantial pressure differentials may exist between the two columns, especially when fluids are being changed or displaced from the hole.

Each length of annulus or tubing containing fluid with a different density is calculated separately. For example, if the annulus has three fluids with densities of d_1, d_2, and d_3, respectively, and the true vertical lengths of coverage for each fluid are h_1, h_2, and h_3, respectively, then the bottomhole pressure in the annulus (P_{an}) is:

$$P_{an} = [d_1 \times h_1) + (d_2 \times h_2) + (d_3 \times h_3)]$$

P_{an} = hydrostatic pressure, psi
d_n = density of fluid in annulus, lb/gal
h_n = true vertical length of coverage of fluid *n* in annulus, ft

If the tubing of length (h_t) is filled with a single fluid of density (d_4) then the equation simplifies to:

$$P_t = d4\ h_t \times 0.052$$

In this case, the pressure differential (P_{dif}) between the annulus and tubing (P_t) is the difference between the pressure exerted by the two columns of fluid:

$$P_{dif} = P_t - P_{an}$$

1.46 Productivity Index

The *productivity index* is s measure of the ability of a well to produce crude oil. And is the ratio of the total liquid flow rate to the pressure drawdown. For a water-free oil, the productivity index (*J*) is given by:

$$J = Q_o/(p_r - p_{wf}) = Q_o/\Delta p$$

In this equation, $Q_o/$ is the oil flow, STB/day, p_r is the volumetric average drainage area pressure (static pressure), p_{wf} is bottom hole flowing pressure, and Δp is the drawdown (psi).

The productivity index for a gas well can be written in a manner analogous to the productivity index for an oil well. Thus:

$$(Q_g)_{max} = J\psi_r$$

Q_g is the gas flow rate, J is the productivity index, and ψ_r is the average reservoir real gas pseudo-pressure, psi²/cp.

1.47 PVT Properties

The specific correlations that should be used for a specific crude oil or reservoir may vary.

1.47.1 Specific Gravity and Molecular Weight

$$\gamma_o = 141.4/(\gamma_{API} + 131.5)$$

$$Mo = (Kw\ \gamma_o^{0.84573}/4.5579)^{6.58848}$$

1.47.2 Isothermal Compressibility

$$X = R_s^{0.1982} T^{0.6685} \gamma_g^{-0.21435} \gamma_{API}^{1.0116} p^{-0.1616}$$

1.47.3 Undersaturated Oil Formation Volume Factor

$$B_o = B_{ob}e^{[co)pb\text{-}p)]}$$

1.47.4 Oil Density

$$p_o = (62.42796\gamma_o + 0.0136\gamma_g Rs)/B_o$$

1.47.5 Dead Oil Viscosity

$$\mu_{od} = [(3.141 \times 10^{10})/T^{3.444}]\log(\gamma_{API})^{[10.313\log(T) - 36.447]}$$

1.47.6 Undersaturated Oil Viscosity

$$\mu_o = \mu_{ob}(p/p_b)^{[2.6p1.18710(-3.9 \times 10-5p-50)]}$$

1.47.7 Gas/Oil Interfacial Tension

$$o_{od} = (1.17013 - 1.694 \times 10^{-3}T)(38.085 - 0.259\gamma_{API})$$

1.47.8 Water/Oil Interfacial Tension

$$T_{co} = 24.2787K_w 1.76544\gamma_o^{2.12504}$$

Calculate the pseudocritical temperature of the gas:

$$T_{cg} = 1.682 + 349.5\gamma_{ghc} - 74.0\gamma_{ghc}^{2}$$

Calculate the pseudocritical temperature of the live gas/oil mixture.

$$T_{cm} - \chi_o T_{co} + \chi_g T_{cg}$$

Convert oil density units from lbm/ft^3 to g/cm^3

$$\rho_h = \rho_o/62.42796 = 0.7206\ g/cm_3$$

Calculate the surface tension between the oil and water phases.

$$\sigma_{hw} = \{[1.58(\rho_w^{'} - \rho_h) + 1.76)/T_r^{0.03125}]\}^4$$

Symbols

B_g = gas FVF, ft^3/scf
B_o = oil FVF, bbl/STB

B_{ob} = oil formation volume at bubble point pressure, bbl/STB
c_o = oil isothermal compressibility, Lt^2/m, psi^{-1}
c_{ob} = oil isothermal compressibility at bubble point, Lt^2/m, psi^{-1}
K_w = Watson characterization factor, $°R^{1/3}$
M_g = gas molecular weight, m, lbm/lbm mol
M_{go} = gas/oil mixture molecular weight, m, lbm/lbm mol
M_o = oil molecular weight, m, lbm/lbm mol
M_{og} = oil-gas mixture molecular weight, m, lbm/lbm mol
p = pressure, m/Lt^2, psia
p_b = bubble point pressure, m/Lt^2, psia
p_bN_2 = bubble point pressure of oil with N_2 present in surface gas, m/Lt^2, psia
p_{bh} = bubble point pressure of oil without nonhydrocarbons, m/Lt^2, psia
p_f = bubble point pressure factor, psia/°R
p_r = pressure ratio (fraction of bubble point pressure)
R_s = solution GOR, scf/STB
T = temperature, T, °F
T_{abs} = temperature, T, °R
T_b = mean average boiling point temperature, T, °R
T_{cg} = gas pseudocritical temperature, T, °R
T_{cm} = mixture pseudocritical temperature, T, °R
T_{co} = oil pseudocritical temperature, T, °R
T_r = reduced temperature, T
T_{sc} = temperature at standard conditions, T, °F
V = volume, L^3
V_o = volume of crude oil, L^3
W_g = weight of dissolved gas, m
W_o = weight of crude oil, m
x_g = gas "component" mole fraction in oil
x_o = oil "component" mole fraction in oil
y_g = gas "component" mole fraction in gas
y_{N_2} = mole fraction N_2 in surface gas
yo = oil "component" mole fraction in gas
Z = gas compressibility factor
γ_{API} = oil API gravity
γ_g = gas specific gravity, air=1
γ_{gc} = gas specific gravity adjusted for separator conditions, air=1
γ_{ghc} = gas specific gravity of hydrocarbon components in a gas mixture, air=1
γ_{gs} = separator gas specific gravity, air=1

γ_o = oil specific gravity
μ_o = oil viscosity, m/Lt, cp
μ_{ob} = bubble point oil viscosity, m/Lt, cp
μ_{od} = dead oil viscosity, m/Lt, cp
ρ_g = gas density, m/L^3, lbm/ft^3
ρ_o = oil density, m/L^3, lbm/ft^3
ρ_{ob} = bubble point oil density, m/L^3, lbm/ft^3
ρ_w = water density, m/L^3, g/cm^3
σ_{hw} = hydrocarbon/water surface tension, m/t^2, dynes/cm
σ_{go} = gas/oil surface tension, m/t^2, dynes/cm
σ_{od} = dead oil surface tension, m/t^2, dynes/cm

1.48 Reserves Estimation

To determine the volume of oil or gas present in the reservoir, the bulk volume of the reservoir (V_b) is first determined using the available reservoir description data. The volume of fluids in the pore spaces of the reservoir rock is then calculated by multiplying the bulk volume by the rock porosity (φ); this is also known as the pore volume of the rock.

The pore volume is normally occupied by oil (or gas) and water. The fraction of the pore volume occupied by water is known as the water saturation (S_w). The porosity and initial water saturation are determined from the logs and core samples obtained from the exploratory wells. Therefore, the initial volume of oil (V_o) at the reservoir conditions is determined by:

$$V_o = V_b \varphi (1 - S_w)$$

This volume of oil is called the initial oil in place (IOIP) or the original oil in place (OOIP).

It is impossible to recover all of the original oil in place – certain forces within the reservoir rock prevent the movement of some oil from the rock to the well. The fraction of the original oil in place that could be recovered is called the recovery factor ($E\gamma$) and the total recoverable volume of oil ($E\gamma V_o$) is called the proven reserves.

1.49 Reservoir Pressure

The discovery pressure of the fluid in a reservoir is related to the weight of its overburden and how much of this is supported by the reservoir rock. The overburden weight is expressed through a pressure that represents the total vertical stress caused by the combined weight of the fluid and rock. The overburden pressure is also called the lithostatic, geostatic, earth, or rock pressure. The support from the reservoir is expressed through a grain pressure, the load carried by grain-to-grain contacts. It is the normal component of the effective stress. Based on a static force balance, the overburden pressure P_o is equal to the sum of the fluid or pore pressure P_p and the grain pressure P_g:

$$P_o = P_p + P_g$$

The overburden pressure is expressed as a pressure gradient, i.e., its change in pressure with respect to depth. The overburden pressure gradient is directly related to the bulk density of the reservoir and overburden materials. The bulk density is, in turn, a function of the grain (rock) density, fluid (water) density, and porosity. The bulk density ρ_b is given by:

$$\rho_b = \rho_w \phi + (1 - \phi)\rho_r$$

where ρ_w and ρ_r are the water and rock (grain) densities, respectively. This equation assumes water is the only interstitial fluid. The bulk density is related to the overburden pressure:

$$p_o = \rho_b g D$$

D is the reservoir subsurface vertical depth, *g* is the gravitational acceleration, and p_o is a relative (gauge) pressure. The product $p_b g$ is the *specific weight,* the force per unit volume or pressure per unit length (i.e., pressure gradient).

The grain density of virtually all sedimentary rocks is approximately equal to 2.65 g/cm^3 (±3 to 5). Water density is approximately 1 g/cm^3. These densities correspond to the following specific weights: $\rho_r g = 1.15$ and $\rho_w g = 0.433$ psi/ft, respectively. Thus, the overburden pressure gradient is generally between 0.433 (if the overburden were fluid with no rock) and 1.15 psi/ft (if there were no water)- a typical value is 1.00 psi/ft.

Pore pressure gradients at discovery are between the hydrostatic and overburden pressure gradients. The hydrostatic pressure gradient is the pressure gradient of the fluids. If the fluid column contains only water, the hydrostatic pressure gradient is 0.433 psi/ft. Because of the effects of dissolved solids in the water, the hydrostatic pressure gradient may be as large as 0.54 psi/ft.

1.50 Resource Estimation

The oil and gas volume/quantities can be estimated by volumetric method. Volumetric oil *in place* in million metric ton is given by the relation:

$$AH\theta(1 - s_w)\rho_o/b_o$$

Where A is the area of oil pool in square kilometers, H is the oil pay thickness in meters, θ is the porosity of the reservoir rocks, ρ_o is the density of oil, b_o is the volume fraction of oil in the formation, and s_w is the fraction saturated by water in the pores

For gas *in place* following relation is used:

$$AH\theta(1 - s_w)p_rT_r/(Zp_sT_s)$$

Where A is the area of oil pool in square kilometer, H is the oil pay thickness in meters, θ is the porosity of the reservoir rocks, s_w is the fraction saturated by water in the pores, p_r is the reservoir pressure in the formation, p_s is the pressure at the surface of earth, T_r is the: absolute temperature in the formation, and T_s is the absolute temperature at the surface.

1.51 Reynold's Number

The Reynolds number, N_{Re}, is a dimensionless number that relates inertial and viscous forces. It is used in the friction factor correlation, to determine the resistance to flow by a pipe.

$$N_{Re} = \rho\ D\ U\ \mu$$

D is the pipe diameter in meters or feet, U is the average fluid velocity in meters/second or feet/second and is equal to G/ ρ, μ is the dynamic viscosity of the fluid in kilogram/meter-second or pounds/foot-hour, ρ is the density of liquid in kilograms/cubic meter or pounds/cubic foot.

1.52 Saturated Steam

The term steam is an imprecise designation because it refers to a water liquid/gas system that can exist from 0°C (32°F) to any higher temperature;

from 0.1 psia to any higher pressure; and from nearly all liquid to 100% gas. Steam quality, f_s, refers to the phase change region of liquid to gas and is defined as:

$$f_s = m_v/(m_v + m_l)$$

In this equation, m_v is the amount of water in the vapor phase and m_l is the amount of water in the liquid phase.

Heat capacity is expressed in units of Btu/(lbm-°F). A "Btu" is defined as the amount of heat required to raise 1 lbm of water from 60 to 61°F. All liquids and solids are compared to pure water, which has the highest heat capacity of any substance at 1 Btu/(lbm-°F). By calculating a ratio of the heat capacity of water divided by that of another substance, a convenient fraction called "specific heat" is obtained. Notice that petroleum has a specific heat of 0.5, or half that of water, and sandstone is only 20% of water on a per pound basis. No other liquid or gas carries as much heat per pound as water. Also, the temperature range at which this high heat carrying performance is achieved, 34 to 700°F, is ideal for many processes, including steam flooding.

Enthalpy is a useful property defined by an arbitrary combination of other properties and is not a true form of energy. Although the absolute value of enthalpy may be of little value, changes in enthalpy are extremely useful, however, and are the basis for steamflood energy calculations. The total enthalpy held by each pound of liquid water at any temperature is called sensible heat, h_f. The heat input, which produces a change of state from liquid to gas without a change of temperature (*latent heat of evaporation*) and is shown by h_{fv}. The total heat, h_v, in each pound of 100% quality or saturated steam is the sum of these two, $h_v = h_f + h_{fv}$.

1.53 Standard Oilfield Measurements

Measurement	Symbol	Units	Other
Capacity (hole, casing, pipe, annulus)	C	bbl/ft	
Density, (SI) kilograms per cubic meter	d	kg/m^3	
Density, pounds per gallon	d	lb/gal	ppg
Diameter, inside	ID	in	
Diameter, outside	OD	in	
Fluid displacement (hole, casing, pipe, annulus)	Dis	ft/bbl	

Measurement	Symbol	Units	Other
Force, pounds	f	lb	lbf
Length, feet	h	ft	
Pi (π), unitless		3.1416	
Pressure, pounds per square inch	P	lb/in^2, psi	
Specific Gravity, unitless	SG	SG	sg, Sp Gr
Velocity, feet per minute	Vel	ft/min	
Volume, barrels	v	bbl	
Volume, cubic feet	v	ft^3	cu. ft.
Volume, cubic inches	v	in^3	cu. in.
Volume, U.S. gallons	v	gal	

1.54 Twist

When a rod is subjected to torque, it undergoes twist, which is expressed by the equation:

$$\theta = TL/Gj \text{ radians}$$

θ is the angle of twist in radians, L is the length of the section in feet, T is the torque, G is the modulus of rigidity in psi and J is the polar motion of inertia.

Example calculation
Consider a pipe with the following dimensions carrying an applied tensile load of 5,000 lbs at the bottom. The pipe outside diameter = 5 in, the pipe inside diameter = 4 in, the pipe density = 490 lb/ft^3, and the pipe length = 30 ft. Calculate the maximum stress in the string.

Solution:
The cross-sectional area of the pipe is:

$$A = \pi/4(5^2 - 4^2) = 7.08 \text{ inches}^2$$

Weight of the pipe = $\pi/4(5^2 - 4^2) \times 490 \times 30 \times 12 = 721.6$ lbf
Total force acting on the top of the pipe:

$$F = \text{weight of pipe} + \text{load applied}$$

$$F = 721.6 + 5000 = 5721.6$$

Maximum stress at the top of the pipe = σ = F/A = 5721.6/7.08 – 809 psi

1.55 Ultimate Tensile Strength

The ultimate tensile strength (UTS) of a material in tension, compression, or shear, respectively, is the maximum tensile, compressive, or shear stress resistance to fracture or rupture. The ultimate tensile strength is equivalent to the maximum load that can be applied over the cross-sectional area on which the load is applied. The ultimate tensile strength of several API pipes is presented in Table 1.16 (below).

1.56 Volume Flow Rate

The volume flow rate, Q, of a fluid is the volume of fluid that is passing through a given cross sectional area per unit time. The term *cross sectional area* refers to the area through which the fluid is flowing, such as the circular area inside a pipe. Thus:

$$Q = V/t$$

Table 1.16 API pipe properties.

AFI	Yield stress, psi		Minimum ultimate	Minimum
Grade	**Minimum**	**Maximum**	**Tensile, psi**	**Elongation (%)**
H-40	40,000	80,000	60,000	29.5
J-55	55,000	80,000	75,000	24
K-55	55,000	80,000	95,000	19.5
N-80	80,000	110,000	100,000	18.5
L-80	80,000	95,000	95,000	19.5
C-90	90,000	105,000	100,000	18.5
C-95	95,000	110,000	105,000	18.5
T-95	95,000	110,000	105,000	18
P-110	110,000	140,000	125,000	15
Q-125	125,000	150,000	135,000	18

V is the volume of fluid in a given time, t. In SI units (International System of Units), the volume flow rate has units of cubic meters per second (m^3/s).

1.57 Volumetric Factors

The formation volume factor of a natural gas (B_g) relates the volume of 1 lb mol of gas at reservoir conditions to the volume of the same lb mol of gas at standard conditions:

$$B_g = \text{(Volume of 1 lb mole of gas at reservoir conditions, RCF)/(Volume of 1 lb mole of gas at standard conditions, SCF)}$$

The formation volume factor of a crude oil or gas condensate (B_o) relates the volume of 1 lb mol of liquid at reservoir conditions to the volume of that liquid once it has gone through the surface separation facility. Thus:

$$B_o = \text{(Volume of 1 lb mole of liquid at reservoir conditions, RB)/(Volume of that 1 lb mole of liquid after going through separation, STB)}$$

The total volume occupied by 1 lb mol of liquid at reservoir conditions $(V_o)_{res}$ can be calculated through the compressibility factor of that liquid:

$$(V_o)_{res} = [(nZ_oRT)/p]_{res}$$

where n = 1 lb mol

Upon separation, some gas is going to be taken out of the liquid stream feeding the surface facility. If n_{st} is the number of the moles of liquid leaving the stock tank per mole of feed entering the separation facility, the volume that 1 lb mol of reservoir liquid is going to occupy after going through the separation facility is:

$$(V_o)_{res} = [(n_{st}Z_oRT)/P]_{sc}$$

This assumes that at the last stage of separation, the stock tank, operates at standard conditions. Thus:

$$Bo = [(nZ_oRT)/P)]/\ [(n_{st}Z_oRT)/P)_{sc}]$$

or,

$$Bo = (1/n_{st})[(Z_o)_{res}/(Z_o)_{sc}](T/P)(P_{sc}/T_{sc}\ [RB/STB]$$

$(Z_o)_{sc}$, unlike Z_{sc} for a gas, is never equal to one.

1.58 Yield Point

The Yield Point (YP) is resistance of initial flow of fluid or the stress required in order to move the fluid and is the attractive force among colloidal particles in drilling fluid. As per Bingham plastic model, the yield point is the shear stress at zero shear rate, and is measured in the field by either:

$$YP = 300 \text{ rpm reading} - PV$$

$$YP = (2 \times 300 \text{ rpm reading}) - 600 \text{ rpm reading}$$

In the equation, PV is the plastic viscosity in centipoise.
The Bingham plastic model is defined by the relationship:

$$\text{Shear Stress} = \text{Yield Stress} + (\text{Plastic Viscosity} \times \text{Shear Rate})$$

Also:

$$\text{Shear Stress} = k \times \text{Shear Rate}_n$$

The constant "k" is the "consistency index" which is indicative of the pumpability of the fluid; "n" is the power index, which denotes the degree of the "non-Newtonian" character of the fluid.

2

RIG Equipment

2.01 API Casing Grades

The API grade of casing denotes the steel properties of the casing. The grade (Table 2.1) has a letter, which designates the grade, and a number, which designates the minimum yield strength in thousands of psi.

Casing properties are defined as:

- Yield Strength: The tensile stress required to produce a total elongation of 0.5% per unit length.
- Collapse Strength: The maximum external pressure or force required to collapse the casing joint
- Burst Strength: The maximum internal pressure required to cause a casing joint to yield

Casing dimensions are specified by its outside diameter (OD) and nominal wall thickness. Normal wellsite conventions specify casing by its OD and weight per foot.

Table 2.1 API Casing Grades.

API Grade	Yield strength (min), psi	Tensile strength (min), psi
H-40	40.000	60.000
J-55	55.000	75.000
K-55	55.000	95.000
C-75	75.000	95.000
L-SO	80.000	100.000
N-80	80.000	100.000
C-90	90.000	105.000
C-95	95.000	105.000
P-110	110.000	125.000

2.02 Block Efficiency Factor

The overall block efficiency factor, E, is avaibale though the relationship:

$$E = (\mu^n - 1)/[\mu^s n(\mu - 1)]$$

In this equation, μ is the friction factor (typically on the order of 1.04), *n* is the number of sheaves (usually $s = n$). Thus a simplified block efficiency factor can be estimate from:

$$E = 0.9787^n$$

2.03 Blocks and Drilling Line

The efficiency of a block and tackle system is measured by:

$$\eta = (\text{power output})/(\text{power input}) = Po/Pi = (F_h v_{tb})/F_f v_f$$

The output power is obtained from the relationship:

$$P_o = F_h v_{tb}$$

F_h is the load hoisted in pounds (buoyed weight of string plus traveling block plus compensator) and v_{tb} is the velocity of the traveling block.

The input power from the draw works to the fast line is given by the relationship:

$$P_i = F_f v_f$$

F_f is the load in fast line, and v_f is the fast line speed. The relationship between the travelling block speed and the fast line speed is:

$$v_{tb} = v_f/n$$

2.04 Crown Block Capacity

The crown block is the stationary section of a block and tackle that contains a set of pulleys or sheaves through which the drill line (wire rope) is threaded or reeved and is opposite and above the traveling block. The combination of the traveling block, crown block and wire rope drill line gives the ability to lift weights in the hundreds of thousands of pounds. On larger drilling rigs, when raising and lowering the derrick, line tensions over a million pounds are not unusual.

The crown block capacity required to handle the net static hook-load capacity can be calculated using the following formula:

$$R_c = [(H_L + S)(n + 2)]/n$$

where R_c is the required crown block rating (lbs), H_L is the net static hook-load capacity (lbs), S is the effective weight of suspended equipment (lbs), and n is the number of lines strung to the traveling block.

Example calculation
What minimum draw-works horsepower is required to drill a well using 10,000 ft of 4.5-inch OD 16.6# drill pipe and 50,000 lbs of drill collars? Efficiency is 65%.

Solution:

Drill string weight (air weight) = 50,000 lbs + (10,000 ft) (16.6 lb/ft) = 216,000 lbs.

Hook horsepower = [(216,000 lbs)(100 ft/min)]/33,000 = 655 hp.

Therefore, the required minimum draw-works horsepower rating = 655/0.65 – 1,007 hp

Please note that this neglects the effect of buoyancy and the weight of the block and hook.

2.05 Derrick Load

The static derrick load is calculated from the relationship:

$$F_s = [(n + 2)/n] \times F_h$$

Dynamic fast line load is calculated from the relationship:

$$F_f = F_h/En$$

The dynamic derrick load is given by the relationship:

$$F_f + F_h + F_{dl}$$

The derrick load is calculated from the relationship:

$$F_d = [(1 + E + En)/En] \times F_h$$

The maximum equivalent derrick load is given as
$F_{de} = [(n + 4)/n] \times F_h$
The derrick efficiency factor is calculated from the relationship:

$$Ed = Fd/F_{de} = [E(n + 1) + 1]/[E(n + 4)]$$

where F_{dl} = the dead line load.

Example calculation
A rotary rig that handles triples is equipped with a 1200 hp draw-works. The efficiency of the hoisting system is 81%. Determine the time it takes to trip one stand of pipe at a hook load of 300,000 pounds.

Solution
The efficiency of the system is:

$$\eta = (\text{power output})/(\text{power input}) = P_o/P_i = 0.81 = P_o/1200 \text{ and } P_o = 927 \text{ hp}$$

Since $P_o = v_{tb}/F_h$

$$v_{tb} = P_o/F_h = [927 \times 33{,}000 \text{ (ft-lb/min)/hp}]/300{,}000 = 102 \text{ fpm}$$

and

$$t = \text{length}/\ v_{tb} = 90 \text{ ft}/102 \text{ ft/min} = 0.84 \text{ min} = 50.5 \text{ sec}$$

2.06 Energy Transfer

Efficiency transfer from the diesel engines to the mud pump can be given as in the Figure 2.1.

Due to interrelated equipment, various efficiencies can be used: (1) engine efficiency, η_e, (2) electric motor efficiency, η_{el}, (3) mud pump mechanical efficiency, η_m, and mud pump volumetric efficiency, η_v from which the overall efficiency is:

$$\eta_o = \eta_e \times \eta_{el} \times \eta_m \times \eta_v$$

Engine input (fuel energy) (1) engine output, i.e. input to pump, (2) mechanical pump output, i.e. input to hydraulic, and (3) hydraulic output

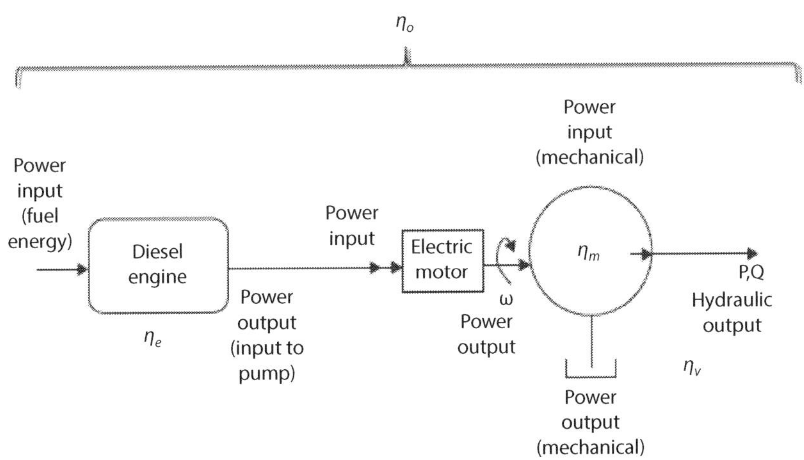

Figure 2.1 Energy Transfer.

Example calculation

Estimate the liner size required for a double-acting duplex pump given the following pump details: (1) rod diameter = 2.5 inches, (2) stroke length = 20-inch stroke, and (3) pump speed = 60 strokes/min. The maximum available pump hydraulic horsepower is 1,000 hp. For optimum hydraulics, the pump recommended delivery pressure is 3,500 psi.

Solution:

The theoretical pump displacement for a duplex pump is:

$$V_i = \pi/4 \times N_c L_s (2D_L^2 - D_R^2) = [2 \times 20(2D_L^2 - 2.52)]/294 \text{ gal/stroke}$$

The theoretical flow rate of the pump operating at 60 strokes/min is:

$$V_t = [2 \times 20(2D_L^2 - 2.5^2)]/294 \times 60 \text{ gal/stroke} = (16.33S_L^2 - 51.0^2) \text{ gpm}$$

The volumetric relationship is:

$$Q_a = Q_t n_\upsilon$$

$$490 = (16.33D_L^2 - 51.02) \times 0.9 \text{ gpm}$$

$$D_L = 6.03 \text{ inches}$$

Therefore, the liner size that can be used is 6 inches.

2.07 Engine Efficiency

The overall efficiency of power generating systems can be calculated from the following relationship:

$$\text{Efficiency (\%)} = (\text{output power})/(\text{input power}) \times 100$$

or

$$n_o = 100(P_v/P_{i)}$$

The output power of an engine is:

$$P_o = (2\pi NT)/33000$$

T = output torque in ft-lbs, N = engine rotary speed in revolution per minute (rpm), and $P\,o$ = output power in horsepower, hp. The input power is expressed as:

$$P_i = (Q_f H)/2545$$

Q_f = rate of fuel consumption in lbm/hr, H = fuel heating value in BTU/lb, and i = input power in horsepower, hp. The fuel consumption can be calculated from:

$$Q_f = 48.46(NT/(\eta H)\ \text{lb/hr}$$

2.08 Line Pull Efficiency Factor

Hoisting engines should have a horsepower rating for intermit-tent service equal to the required draw-works horsepower rating divided by 85% efficiency. Draw works also have a line pull rating efficiency depending on the number of lines strung between crown block and traveling block (Table 2.2).

Example calculation
What line pull is required to handle a 500,000 lb casing load with 10 lines strung?

Solution

$$\text{Line pull} = 500{,}000/(10)\ (0.81) = 61{,}728\ \text{pounds}$$

2.09 Mud Pump

A mud pump is a large reciprocating pump used to circulate the mud (drilling fluid) on a drilling rig. The pump uses a reciprocating piston/

Table 2.2 Effect of the number of Lines on Efficiency.

No. of lines	Efficiency factor
6	0,074
8	0,042
10	0.811
12	0,782

plunger device that is designed to circulate drilling fluid under high pressure (up to 7,500 psi, 52,000 kPa)) down the drill string and back up the annulus.

2.09.1 Volume of Fluid Displaced

The following calculation can be used to determine the theoretical volume of fluid displaced:

For a single-acting pump:

$$V_t = [(\pi/4)D_L^2L_s)]N_c$$

For a double-acting pump:

$$V_t = (\pi/4)NcL_s)(2D_L^2 - D_r^2)$$

The actual flow rate:

$$Q_a = Q_t\eta_v$$

The pump hydraulic horsepower is:

$$HHP_p = (P_pQ)/1714$$

D_L is the liner or piston diameter in inches, L_s is the stroke length in inches, N_c is the number of cylinders, 2 for duplex and 3 for triplex, D_r is the rod diameter in inches, and h_v is the volumetric efficiency.

2.09.2 Volumetric Efficiency

The volumetric efficiency is determined using Qa, the actual flow speed:

$$\eta_v = Q_a/(\text{displacement volume} \times \text{speed}) \times 100$$

Thus:

$$\eta_v = (Q_t - \Delta Q)/Q_t \times 100$$

ΔQ is leakage losses and Q_t is the theoretical flow rate which is equal to the displacement volume multiplied by the pump speed.

2.09.3 Pump Factor

The pump factor is the pump displacement per cycle and is given as *PF* in bbl/stroke or gal/stroke.

The duplex pump factor is given by the relationship:

$$2(V_{fs} + V_{bs}) = PF_d$$

$$PF_d = (\pi/2)L_s(2D_L^{\,2} - D_r^{\,2})$$

The triplex pump factor is given by the relationship:

$$3V_{fs} = PF_t$$

$$PF_t = (3\pi/4)D_L^{\,2}L_s$$

The volumetric efficiency is given by the relationship:

$$\eta_v = (PF_a)/(PF_t) \times 100$$

PF_a is the actual pump factor, and PF_t is the theoretical pump factor.

For duplex pumps:

$$V_t = (\pi/4)N_cL_s(2D_L^{\,2} - D_r^{\,2})\eta_v$$

$$V_t = [N_cL_s(2D_L^{\,2} - D_r^{\,2})\ \eta_v]/(42 \times 296) \text{ bbl/stroke}$$

For triplex pumps (single-acting, three cylinders):

$$V_t = [L_s(D^2)\ \eta_v]/(42 \times 98.3) \text{ bbl/stroke}$$

2.10 Offshore Vessels

Offshore vessels are ships that specifically serve operational purposes such as oil exploration and construction work on the oceans. There are a variety of offshore vessels, which not only help in exploration and drilling of oil but also for providing necessary supplies to the excavation and construction units located on the oceans.

2.10.1 Terminology

The sides of offshore vessels and the types of vessel motions encountered are shown in Figure 2.2, Figure 2.3, and Figure 2.4.

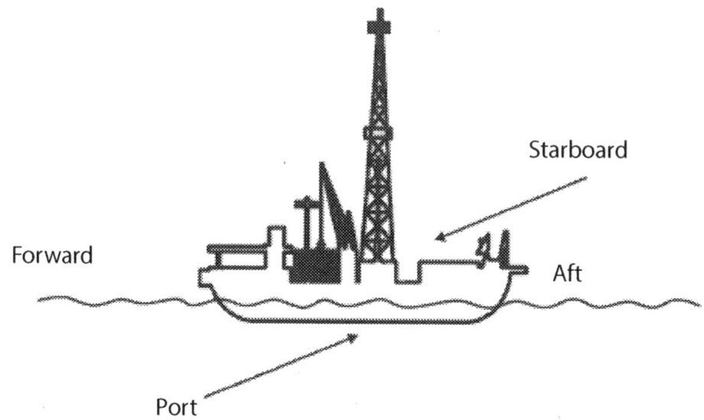

Figure 2.2 Schematic of an Offshore Vessel.

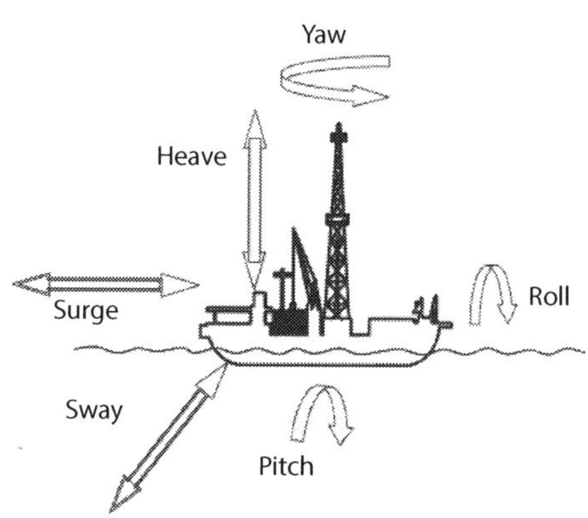

Figure 2.3 Vessel Motion.

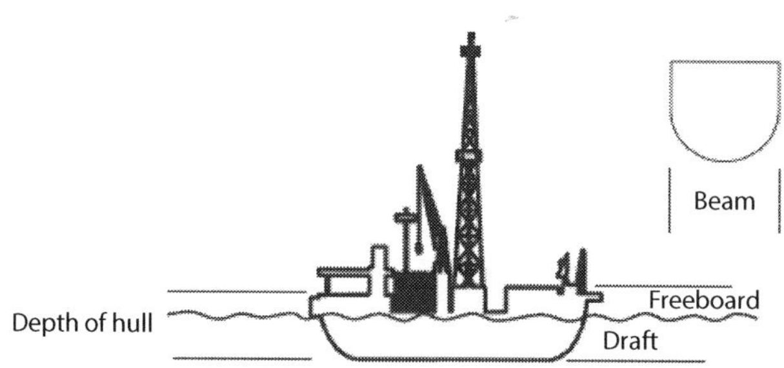

Figure 2.4 Vessel Draft and Freeboard.

In action, the depth of the vessel in the water (draft) and the distance above the water (freeboard) are also important factors.

2.10.2 Environmental Forces

Environmental forces that influence offshore vessels include: (1) wind force, (2) wave force, and (3) current force.

The wind force on an offshore vessel is calculated using the relationship:

$$F_w = 0.00338 V_w^{\,2} C_h C_s A$$

In this equation, F_w is the wind force lbf, V_w is the wind velocity in knots, C_h is the height coefficient, Cs is the shape coefficient, and A is the projected area of all exposed surface in squire feet. The shape coefficients and height coefficients can be estimated from the tables below:

The current force is calculated from the equation:

$$F_c = g_c V_c^{\,2} C_s A$$

F_c is the current drag force lbf, g_c V_c is the velocity of the current, C_s is the drag coefficient and is the same as the wind coefficient, and A is the projected area of all exposed surfaces in square feet.

2.10.3 Riser Angle

The riser angle is measured relative to the vertical. Riser angles are measured with respect to the × and y axis, and the resultant riser angles follow.

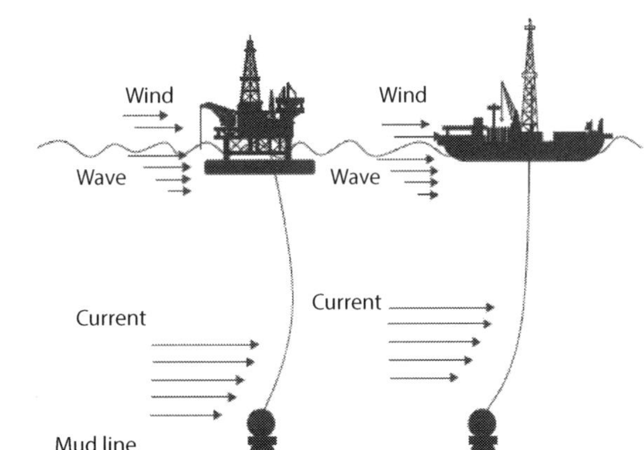

Figure 2.5 Environmental Forces that Influence Offshore Vessels.

Table 2.2 Shape Coefficients.

Shape	Cs
Cylindrical shapes	0.5
Hull (surface type)	1.0
Deck House	1.0
Isolated structural shapes (cranes, beams..etc)	10
Under Deck areas (smooth surfaces)	10
Under Deck areas (exposed beams, girders)	13
Rig derrick (each face)	1.25

Table 2.3 Height Coefficients.

From to (ft)	Ch
0–50	1.0
50–100	1.1
100–150	1.2
150–200	1.3
200–250	1.37
250–300	1.43
300–350	1.48
350–400	1.52
400–450	1.56
450–500	1.6

The equation is:

$$\theta = \tan^{-1}(\tan^2\theta_x + \tan^{2\,\theta y})^{1/2}$$

Example calculation

Calculate the percentage error in using the approximate equation to calculate the resultant riser angle when the riser angle in × and y directions are 4° and 5°, respectively.

Solution:

Using the equation above:

$$\theta = \tan^{-1}(\tan^2\theta_x + \tan^2\theta_y)^{1/2} = \tan^{-1}(\tan^2 4 + \tan^2 5)^{1/2} = 6.39°$$

2.11 Rotary Power

Rotary horsepower can be calculated using the following equation:

$$H_{rp} = (2\pi NT)/33000$$

N is the speed of the rotary table in rpm and T is the torque in ft-lbs.

Example calculation
If a drill pipe has a maximum recommended make-up torque of 20,000 ft-lbs, then what is the rotary horsepower that can be transmitted at 100 rpm?

Solution:
From the equation above:

$$H_{rp} = (2\pi \times 100 \times 20{,}000)/33{,}000 = 381 \text{ hp}$$

The empirical relationship to estimate the rotary horsepower requirements is:

$$H_{rp} = F \times N$$

F is the torque factor in ft-lbf and *N* is the rotary speed in rpm. The torque factor (*F*) is generally estimated as follows: (1) F = 1.5 to 1.75 for shallow holes less than 10,000 ft with light drill string, (2) F = 1.75 to 2.0 for 10,000–15,000 ft wells with average conditions, (3) F = 2.0 to 2.25 for deep holes with heavy drill string, and (4) F = 2.0 to 3.0 for high-torque.

The above empirical estimates are subject to many variables but have proved to be reasonable estimates of rotary requirements. However, for highly deviated wells, torque/H_{rp} requirements must be closely calculated using available computer software programs.

2.12 Ton-Miles Calculations

2.12.1 Round-Trip Ton Miles Calculations

$$Tf = [(L_h + L_s)W_{dp} + 4L_h(Wb + 0.5W_1 + 0.5W_2 + 0.5W_3)]/10{,}560{,}000 \text{ ton-miles}$$

where L_h is the measured depth of the hole or trip depth in feet, L_s is the length of the stand in feet, W_{dp} is the buoyed weight of the drill pipe per foot, ppf, W_b is the weight of the block, hook, etc. in pounds, W_1 is the excess weight of drill collar in mud, lbs (buoyed drill collar weight in mud – buoyed drill pipe weight of same length in mud), W_2 is the excess weight of heavy weight pipe in mud, lbs (buoyed heavy weight pipe in mud – buoyed drill pipe weight of same length in mud), and W_3 is the excess weight of miscellaneous drilling tools in mud, lbs (buoyed miscellaneous drilling tools in mud – buoyed drill pipe weight of same length in mud).

2.12.2 Drilling Ton-Miles Calculations

When a hole is drilled only once without any reaming, the relationship is:

$$T_d = 2(T_{i+1} - T_i)$$

When a hole is drilled with one-time reaming, the relationship is:

$$T_d = 3(T_{i+1} - T_i)$$

When a hole is drilled with two times reaming, the relationship is:

$$T_d = 4(T_{i+1} - T_i)$$

T_i is the round-trip ton-mile calculated from depth i.

2.12.3 Coring Ton-Miles Calculations

$$T_d = 2(T_{i+1} - T_i)$$

2.12.4 Casing Ton-Miles Calculations

$$T_f = 0.5[L_h(L_s + L_h)W_c + 4W_bL_h)/10{,}560{,}000$$

W_c is the buoyed weight of the casing per foot.

Example calculation

Calculate the ton-mile for a round trip from 8,000 feet with 450 feet of drill collar with a weight of 83 ppf. The weight of the drill pipe is 19.5 ppf. The mud density is 9.6 ppg. Assume the stand length to be 93 feet and the total block weight is 40,000 lbs.

Solution:
From the equation above, L_h is 8000 ft, L_s is 90 ft, and W_b is 40000 lbs. The buoyancy factor is 0.853 (i.e., 1 – 9.6/65.5).

$$W_{dp} = 19.5 \times 0.853 = 16.64 \text{ ppf.}$$

The (buoyed weight of drill collar) minus (the buoyed weight of drill pipe) = 0.853 × 450(83 – 19.5) = 24,387 lbs.

Thus:

$$T_f = [8000(93 + 8000)16.64 + 4 \times 8000(40000 + 24387)]/10{,}560{,}000 = 297 \text{ TM.}$$

3

Well Path Design

3.01 Average Curvature-Average Dogleg Severity

The equations used to calculate the average curvature of a survey interval are:

$$k = [(\Delta\alpha/\Delta L)^2 + (\Delta\varphi/\Delta L)^2 \sin^2 \alpha)]^{1/2}$$

$$k = \beta/\Delta L$$

In these equations, α is the inclination (in degrees), φ is the azimuth or direction in degrees, $\bar{\alpha}$ is the average inclination angle (in degrees), k is the average borehole curvature, and β is: $a\cos(\cos\alpha_1\cos\alpha_2 + \sin\alpha_1\sin\alpha_2\cos\Delta\varphi)$.

3.02 Bending Angle

Using the inclination angles and directions, the borehole beding angle is derived from the following equation:

$$\cos\beta = \cos\alpha_1\cos_2 + \sin_1\sin_2\cos\Delta\varphi$$

$\Delta\varphi$ is equal to $\varphi_2 - \varphi_1$, β is the bending angle in degrees, $\Delta\varphi$ is the section increment of azimuth angle in degrees, α_1 is the inclination angle at survey point 1 in degrees, and α_2 is the inclination angle at survey point 2 in degrees.

3.03 Borehole Curvature

3.03.1 General Formula

Borehole curvature is derived by using the general formula:

$$k = (k_\alpha^{\ 2} + k\phi^2 \sin 2\alpha)^{1/2}$$

In addition, the general formula for borehole curvature with the vertical and horizontal curvatures can be is:

$$k = (k_v^{\ 2} + k_H^{\ 2} \sin^4\alpha)^{1/2}$$

κ_v is the curvature of wellbore trajectory in a vertical expansion plot in degrees/30 meters or in degrees/100 feet, and κ_H is the curvature of wellbore trajectory in a horizontal projection plot in degrees/30 meters or in degrees/100 feet.

3.03.2 Borehole Radius of Curvature

The borehole radius of curvature and torsion can be determined by use of either of two formulas:

$$R = (180C_k)/(\pi k)$$

$$\rho = (180C_k)/(\pi\tau)$$

C_κ is the constant related to the unit of borehole curvature. If the unit for borehole curvature is degrees/30 meters and degrees/100 feet, then C_κ is equal to 30 and 100, respectively.

K is the curvature of wellbore trajectory in degrees/30 meters (or degrees/100 feet) and τ is the torsion of wellbore trajectory in degrees/30 meters (or degrees/100 feet).

Example calculation

Calculate the radius of curvature for a buod section with a build rate of 2 degree per 100 feet.

Solution
Using the first of the two equations presented in this sub-section, the radius of curvature is:

$$R = (180C_k)/(\pi k)$$

$$C_k = 100$$

Thus, the radius of curvature, R, is:

$$R = (180 \times 100)/\pi 2 = 2{,}863.63 \text{ feet}$$

3.04 Borehole Torsion

3.04.1 General Method

The borehole bending angle can be derived using the following equation:

$$\tau = (k_\alpha k_\phi - k_\phi k_\alpha / k^2) \times \sin\alpha + k_\phi (1 + k_\alpha^{\ 2}/k^2)\cos\alpha$$

τ is the torsion of wellbore trajectory in degrees/30 meters or in degrees/100 feet, k_α is the first derivative of inclination change rate, viz., the second derivative of inclination angle, and $\kappa\varphi$ is the first derivative of azimuth change rate, viz., the second derivative of azimuth angle.

3.04.2 Cylindrical Helical Method

When the wellbore curvature equals zero, the wellpath is a straight line that will result in zero torsion. When κ is not equal to zero, the torsion equation for the cylindrical helical model is:

$$\tau = k_H[1 + (2k_v^{\ 2})/k^2]\sin\alpha \cos\alpha$$

Example calculation
If the original hole inclination is 22°, to reach the limits of the target, it is desired to build an angle of 26° in a course length of 100 feet and the directional change is –5°. What is the resulting curvature and torsion that will achieve the desired objective?

Solution
Since $\alpha_1 = 22°$; $\alpha_2 = 26°$; $\Delta\varphi = -5°$; $k_\alpha = 4°/100$ ft; $k_\varphi = 7°/100$ ft, and $\Delta L =$ 100 feet, the vertical curvature is:

$$k_\alpha = (\alpha_2 - \alpha_1)/L_2 - L_1) = 4/100 \times 100 = 4° \text{ per 100 feet}$$

The average inclination angle is:

$$\acute{\alpha} = (\alpha_1 + \alpha/_2) = (22 + 26)/2 = 24°$$

$$k_\alpha = k_v = 4° \text{ per 100 feet}$$

$$kH = k\phi/\sin\alpha = -10/\sin 24 = 12.294° \text{ per 100 feet}$$

The wellbore curvature can be calculated as:

$$k = (k_v^{\ 2} + k_H^{\ 2} \sin^4\alpha)^{1/2} = 4.487° \text{ per 100 feet}$$

The torsion is calculated as:

$$\tau = -12.294[1 + (2 \times 4^2)/4.487^2]\sin 24\cos 24 = -11.827° \text{ per 100 feet}$$

3.05 Horizontal Displacement

On the horizontal plane, the displacement of the target point can be calculated from the equation:

$$H_t = [(N_t - N_o)^2 + E_t - E_o)^2]^{1/2}$$

N_t is the distance northing of target (i.e. the refers to the northward-measured distance) in feet, ft, N_o is the distance northing of the slot in feet, E_t is distance easting of target (i.e. the refers to the eastward-measured distance) in feet, and E_o is the distance easting of slot in feet.

The target bearing is given as:

$$\Phi_t = \tan^{-1}(E_t - E_o)/N_t - N_o)$$

Example calculation
With the slot coordinates 1200.5 ft N, 700 ft E and target coordinates 3800 ft N, 4520 ft E, calculate the horizontal displacement of the well. Also, calculate the target bearing.

Solution
Since $N_t = 3800$ ft, $N_o = 1200.5$ ft, $E_t = 4520$ ft, and 700 ft The horizontal displacement is:

$$H_t = [(3800 - 1200.5)^2 + (4520 - 700)^2]^{1/2} = 4620.6 \text{ ft}$$

The target bearing is:

$$\varphi_t = \tan^{-1}[4520 - 700)/3800 - 1200.5)] = 55.76° \text{ or N55.76E}$$

3.06 Magnetic Reference and Interference

Declination (Figure, below) is the angular difference in azimuth readings between magnetic north and true north. Magnetic declination is positive when magnetic north lies east of true north, and it is negative when magnetic north lies west of true north. It is the error between the true north and magnetic north for a specific location.

Azimuth correction can be given as:

$$\text{Azimuth (true)} = \text{Azimuth (magnetic)} + \text{Magnetic declination}$$

For westerly declination, the azimuth correction is:

$$\text{Azimuth (true)} = \text{Azimuth (magnetic)} + (-\text{Magnetic declination})$$

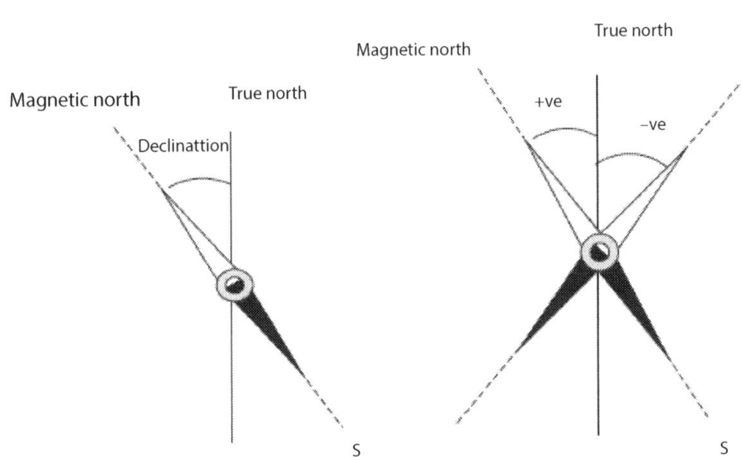

Figure Declination/East-West Declination.

For true North and grid North, true north uses latitude and longitude coordinates of the curved earth as the reference. Longitudes converge upon the rotational pole. In the grid system, the Y-axis does not converge to single point.

Magnetic Declination – Grid Convergence = Total Correction

Magnetic Azimuth + Total Correction = Corrected Azimuth

Example calculation
Determine the azimuth with respect to the true north of the following wells:

1. N45E, declination 5 degrees west,
2. N45E, declination 5 degrees east.

Solution
1. Azimuth = 45 – 5 = 40 degrees = N40 E,
2. Azimuth = 45 + 5 = 50 = N50 E (as shown below).

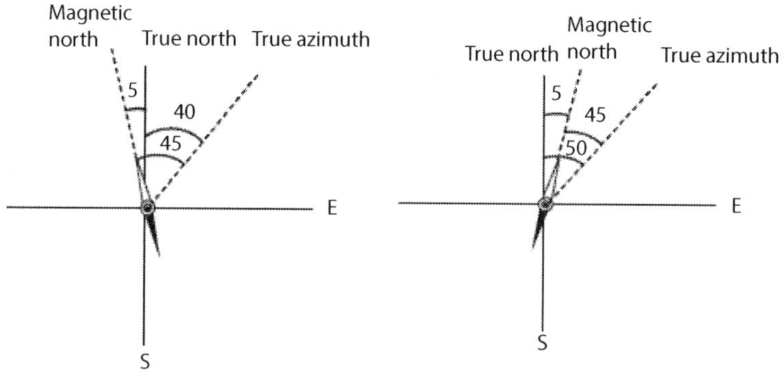

3.07 Tool Face Angle

With the inclination angles and directions, the tool face rotation is derived from the equation:

$$\gamma = arc\ \cos[(\cos \alpha_1 \cos \beta - \cos \alpha_2)/(\sin \alpha_2 \sin \beta)]$$

Example calculation
Calculate the dogleg severity from the following date:

Measured depth (ft)	Angle (degrees)	Direction (degrees)
15000	16.5	150
15045	17.5	148

$$\gamma = \arccos\left[\frac{(\cos 2.01 \cos 5.5 - \cos 7.5)}{(\sin 7.5 \sin 2.01)}\right] = 43.270°.$$

Problem 3.8

While drilling a directional well, a survey shows the original inclination of 32 deg and an azimuth of 112 deg. The inclination and directional curvatures are to be maintained at 4 deg/100 ft and 7 deg/100 ft, respectively, for a course length of 100 ft. Calculate the new inclination angle and final direction. Also, determine the curvature, vertical curvature, and horizontal walk curvatures.

Solution:

Given data: a = 32 deg; <t>$_2$= 112 deg; = 4 deg/100 ft; v =7 deg/100 ft; AL = 100 ft

The new inclination angle is

$$\alpha_2 = \alpha_1 + \kappa_a (L_2 - L_1) = 32 + \frac{4}{100} \times 100 = 36°.$$

The new direction is

$$\phi_2 = \phi_1 + \kappa_\phi (L_2 - L_1) = 112 + \frac{7}{100} \times 100 = 119°.$$

The average inclination angle is

$$\bar{a} = \frac{a_1 + a_2}{2} = \frac{32 + 36}{2} = 34°$$

The average curvature is

$$\kappa = \sqrt{4^2 + 7^2 \sin^2 34} = 5.596° / 100 ft.$$

Using the bending or dogleg angle,

$$\cos\beta = \cos\alpha_1 \cos\alpha_2 + \sin\alpha_1 \sin\alpha_2 \cos\Delta\phi,$$

$$\bar{k} = \frac{arc\cos(\cos 32 \cos 36 + \sin 32 \sin 36 \cos 7)}{100} = 5.591° / 100 ft.$$

It can be seen that a small difference is observed between the two calculation methods.

Vertical curvature is

$$\kappa_v = \kappa_a = 4°/100 ft.$$

Horizontal curvature is

$$\kappa_H = \frac{7}{\sin 34} = 12.518°/100 ft.$$

3.08 Tool Face Angle Change

If the inclination α, new inclination α_n, and the dogleg severity β are known, then the tool face angle can be calculated using the following relationship:

$$\gamma = arc\cos[(\cos\alpha\cos\beta - \cos\alpha_n)/(\sin\alpha\sin\beta)]$$

If the new inclination α_n, azimuth change Δφ, and the dogleg severity *b* are known, then the tool face angle can be calculated using the following relationship:

$$\gamma = arc\sin[(\sin\alpha_n\sin\Delta\beta)/(\sin\beta)]$$

If the inclination α, the dogleg severity β, and the toolface angle *g* are known, the new inclination angle can be calculated using the following relationship:

$$\alpha_n = arc\cos(\cos\alpha\cos\beta - \sin\beta\sin\alpha\cos\gamma)$$

If the inclination α, the dogleg severity β, and the tool face angle γ are known, the change in the azimuth can be calculated using the following relationship:

$$\Delta\varphi = arc\tan[(\tan\beta\sin\gamma)/\sin\alpha + \tan\beta\cos\alpha\cos\gamma)]$$

Example calculation

Derive an equation to find the overall angle change in terms of Δφ, α, and γ, the tool face rotation angle.

Solution
Starting from the azimuth change:

$$\Delta\varphi = arc\tan(\tan\beta\sin\gamma)/\sin\alpha + \tan\beta\cos\alpha\cos\gamma)$$

$$\tan\Delta\varphi = (\tan\beta\sin\gamma)/\sin\alpha + \tan\beta\cos\alpha\cos\gamma)$$

Thus:

$$\tan\Delta\varphi\sin\alpha + \tan\beta(\tan\Delta\varphi\cos\alpha\cos\gamma\text{-}\sin\gamma) = 0$$

This leads to an equation for the overall angle change, β

$$\mathrm{B}\,\beta = arc\tan[(\tan\Delta\varphi\sin\alpha)/(\sin\gamma\text{-}\tan\Delta\varphi\cos\alpha\cos)]$$

3.09 Tortuosity

Tortuosity is a property of a curve being tortuous (twisted; having many turns) and there have been several attempts to quantify this property. Wellbore tortuosity is derived from the equation:

$$T = \frac{\sum_{i=t}^{m} \alpha_{n-1} + \Delta D \times \beta_i}{D_i - D_{i-1}}.$$

3.09.1 Absolute and Relative Tortuosity

The equation is:

$$\Gamma_{(abs)_\pi} = \left(\frac{\sum_{i=1}^{i=1} \alpha_{adj}}{D_n + \Delta D_n} \right),$$

In this equation, $\alpha_{\text{adj}} = a_i + \Delta D_i \times b_i$ and is the dogleg adjusted, summed total inclination angle.

$$\Gamma_{(rel)\pi} = \Gamma^{tar}_{(abs)_\pi} - \Gamma^{notar}_{(abs)\pi}\ \text{deg}/100\,ft.$$

The period of the sine wave should not be of 2/n (n = 1, 2, 3...)

Different methods include (1) the sine wave method, (2) the helical method, (3) random inclination and the azimuth method, and (4) the random inclination dependent azimuth method.

3.09.2 Sine Wave Method

$$\Delta\alpha = \sin(D/P \times 2\pi) \times M$$

D is the measured depth in feet, P is the period, and M is the magnitude. If the measured depth of the survey point is an exact integer multuiple of the period, then:

$$\Delta\alpha = \sin(MD/P \times 2\pi) = 0$$

In addition, the inclination angle is modified so that it does not become less than zero, since negative inclination angles are not allowed.

If the $\Delta\alpha$ term is negative and $\varphi = \varphi_n + \Delta\alpha$ is negative, then $\varphi_n = 360 + \Delta\alpha$, then:

$$\alpha_n = \alpha + \Delta\alpha$$

$$\varphi_n = \varphi + \Delta\alpha + \psi_{xvc}$$

ψ_{xvc} is the cross vertical correction.

If $\alpha_n < 0$, the cross vertical correction = 180° or if $\alpha_n \geq 0$ the cross vertical correction = 0°.

Example calculation

Measured depth = 3900 ft, Inclination = 5.15°, and Azimuth = 166°.

Measured depth = 3725 ft, Inclination = 3.25°, and Azimuth = 165°.

Calculate the tortuosity using the sine wave method.

Solution:

$$\Delta\alpha = \sin(3725/1000 \times 2\pi) \times 1 = -0.99°$$

$$\varphi_n = 165 - 0.99 + 0 = 164.1°$$

$$\alpha_n = 3.25 - 0.99 - 2.26°$$

3.09.3 Helical Method

The helical method modifies the inclination and azimuth of the survey points by superimposing a helix along the wellbore path using the magnitude (radius of the cylinder in the parametric equation) and period (pitch) specified. This method uses the circular helix defined as: $f(u) = a\cos(u) + a\sin(u) + bu$.

The generalized parametric set of equations for the helix used to superimpose the wellbore path is given by:

$$\begin{cases} x(u) = M\cos(u) \\ y(u) = M\sin(u) \\ z(u) = \dfrac{P}{2\pi}u \end{cases}$$

3.09.4 Random Inclination Azimuth Method

$$\alpha_n = \alpha + \Delta\alpha$$

$$\varphi = \varphi_n + \Delta\alpha + \psi_{cvc}$$

As above, ψ_{cvc} is the cross vertical correction.

Example calculation

Measured depth = 3725 ft, Inclination = 3.25°, and Azimuth = 165°.

Measured depth = 3900 ft, Inclination = 5.15°, and Azimuth = 166°.

ζ (Rand number) = 0.375.

Calculate the tortuosity using the random inclination azimuth method.

Solution:

$$\Delta\alpha = [0.375 \times (3900 - 3725/100)] \times 1 = 0.66^{\circ}$$

$$\alpha_n = 5.15 + 0.66 = 5.81^{\circ}$$

$$\varphi_n = 166 + 0.66 + 0 = 166.66^{\circ}$$

3.09.5 Random Inclination Dependent Azimuth Method

$$\Delta\alpha = \zeta \times \delta$$

$$\delta = (\Delta MD/P)M$$

ζ is the random number, $\alpha_n = \alpha + \Delta\alpha$

$$\varphi_n = \varphi + [\Delta\alpha/(2 \sin \alpha_n)] + \psi_{cvc}$$

Example calculation

Measured depth = 3725 ft, inclination = 3.25°, and azimuth = 165°.

Measured depth = 3900 ft, inclination = 5.15°, and azimuth = 166°.

Random number = 0.375.

Calculate the tortuosity using the random inclination dependent azimuth method.

Solution:

$$\Delta\alpha = [0.375 \times (3900 - 3725)]/100 \times 1 = 0.66^{o}$$

$$\alpha_n = 5.15 + 0.66 = 5.81o$$

$$\varphi_n = 166 + (0.66/2 \sin 5.81) + 0 = 169.26^{o}$$

3.10 Types of Wellpath Designs

The different types of Wellpath designs are shown in Figure (below):

3.11 Vertical and Horizontal Curvatures

Vertical and horizontal curvatures can be calculated using the following equations:

$$\kappa_v = \kappa_\alpha$$

$$\kappa_v = d\alpha/dL = \kappa_\alpha$$

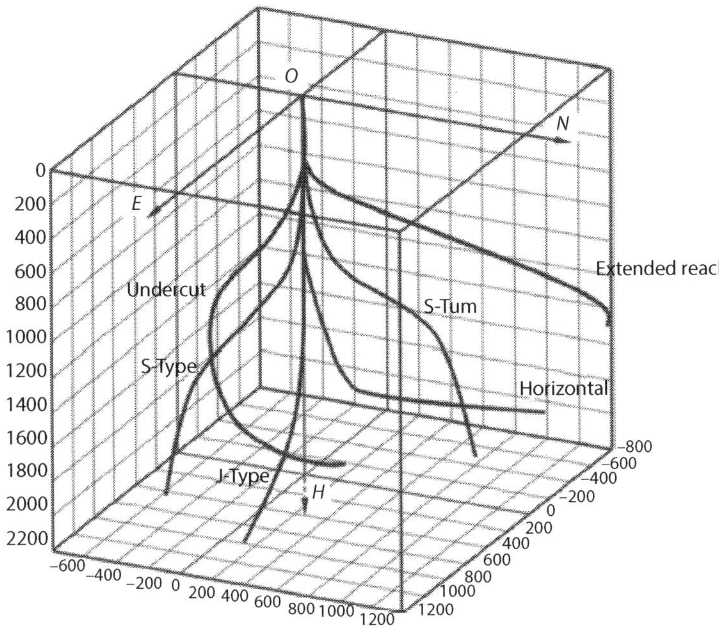

Figure 3.1 Wellpath Designs.

$$\kappa_H = \kappa\varphi/\sin\alpha$$

$$\kappa_H = d\varphi/dS = d\varphi/(dL \sin\alpha) = \kappa_\varphi/\sin\alpha$$

k_V is the curvature of the wellbore trajectory in a vertical position plot in degrees per 30 meters or degrees per/100 feet, k_H is the curvature of the wellbore trajectory in a horizontal projection plot, degrees per 30 meters or degrees per 100 feet, $dL = L_2 - L_1$ and is the curved section length in meters or feet, and S is the arc length in the azimuthal direction in meters or feet.

3.12 Wellbore Trajectory Uncertainty

Nominal distance between two points, i.e., from the point concerned in a reference wellbore (the reference point) to the point observed in the object wellbore (the object point), is:

$$S_{or} = [(x_o - x_r)^2 + (y_o - t_r)^2 + (z_o - z_r)^2]^{1/2}$$

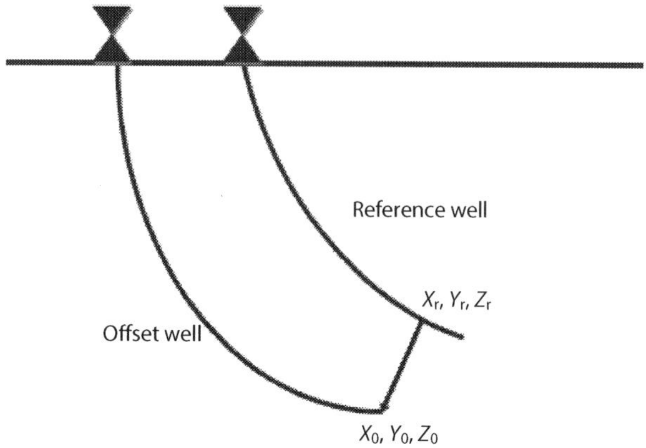

Figure 3.2 Reference and Offset Well.

In this equation, x_o, y_o, and z_o are coordinates of the object point in the OXYZ system and x_r, y_r, and z_r are coordinates of the object point in the OXYZ system S_{or} (Figure).

If rs represents the sum of the radii, then:

$$r_s = r_o + r_r$$

where r_o is the radius of an object well and r_r is the radius of a reference well. The relative covariance of the matrices are:

$$\Sigma_{RR} = \Sigma_{PoPo} + \Sigma_{RrPr}$$

Σ_{PoPo} and Σ_{RrPr} are the covaroaince matricies of the points P_o ands P_r, respectively. The azimuth and dip angle of the single point can be determined by:

$$\varphi = \text{arc sin } [U_r^{\ 2} + V_r^{\ 2})/S_{or}] \text{ and } \theta = \text{arc tan } V_r/U_r$$

If k_p is the amplifying factor corresponding to an ellipsoid on which the signal point lies, it is given by:

$$k = DS/(\sigma_1^{\ 2}\cos^2\theta\sin^2\phi + \sigma_2^{\ 2}\sin^2\theta\sin^2\phi + \sigma_3^{\ 2}\cos^2\phi$$

In this equation, ϕ is the azimuth angle and θ is the dip angle.

3.13 Wellpath Length Calculations

Circular arc:

$$\text{Arc length} = L_c = (\alpha_1 - \alpha_2)\text{BRA}$$

BRA is the build rate angle in degrees per 100feet.
Vertical distance is given by:

$$V - R_b(\sin\alpha_2 - \sin\alpha_1)$$

Rb is the radius of the build or:

$$Rb = 180/(\pi \times \text{BRA})$$

Horizontal distance, H, is given by:

$$H = R_b(\cos\alpha_1 \text{-} \cos\alpha_2)$$

Tangent section:

Length of tangent = L_t
Vertical distance = $V = L_t \cos\alpha$
Horizontal distance = $H = L_t \sin\alpha$

Example calculation
Determine the build rate radius from the data for a build-and-hold pattern type well: build rate angle = 30 degrees per w00 feet.

Solution

$$R_b = 1/\text{BRA}(180/\pi) = 100/3(180/\pi) = 1910 \text{ feet.}$$

4

Fluids

4.01 Acidity-Alkalinity

Acidity or alkalinity is expressed as the pH of the material or solution (Table 4.1).

The pH is the negative logarithm of $[H^+]$ or $[OH^-]$ and is a measurement of the acidity or alkalinity of a solution:

$$pH = -\log([H^+])$$

$$pH = -\log([OH^-)]$$

$[H^+]$ or $[OH^-]$ are hydrogen and hydroxide ion concentrations, respectively, in moles/liter. Also, at room temperature, $pH + pOH = 14$. For other temperatures: pH + pOH = pKw, where Kw = the ion product constant at that particular temperature.

Table 4.1 Ranges of Acidity and Alkalinity.

pH	[H–]	Solution
<7	$> 1.0 \times 10^{-7}$ M	Acid
>7	$< 1.0 \times 10^{-7}$ M	Basic
7	$= 1.0 \times 10^{-7}$ M	Neutral

Example calculation
An aqueous potassium hydroxide completion fluid has a pH of 9. Determine the hydrogen ion concentration of this solution?

Solution:
From the pH equation (abpove):

$$pH = -\log([H^+])$$

The hydrogen ion concentration is:

$$[H^+] = 10^{-pH} = 10^{-10} = 1 \times 10^{-9} \text{ mol/L } (10^{-9} \text{ M}).$$

4.02 Base Fluid–Water-Oil Ratios

The base fluid/water ratio can be calculated as the percentage by volume of base fluid in a liquid phase:

$$P_b = VR_b/(VR_b + V\,R_w) \times 100$$

VR_b is the volume percent of base fluid and VR_w is the volume percent of water. The percent of water is $P_w = 100 - P_b$

Example calculation
If the base fluid is 62%, water is 12%, and solids is 26%, calculate the base fluid-water ratio.

$$P_b = VR_b/(VR_b + V\,R_w) \times 100 = 62/(62 + 12) \times 100 = 83.78$$

$$P_w = 100 - P_b = 100 - 83.78 = 16.22$$

$$P_b/P_w = VR_b/VR_w = 83.78/16.22 = 5.17$$

4.03 Common Weighting Materials

A weighting material is a high-specific gravity and finely divided solid material that is used to increase the density of a drilling fluid. Barite is the most common weighting material as is hematite (the mineral form of ferric oxide, Fe_2O_3). Calcium carbonate, specific gravity is also a weighting material but is used more for its acid solubility than for density modification.

The average weights of the commonly used weighting materials are given in the Table 4.2 (below). Table 4.3 and Table 4.4 (below) provide general size range (in microns) of a variety of solids.

Table 4.2 Common Weighting Materials.

Weighting materials	Specific gravity (gm/cm3)	Density (lbm/gal – ppg)	Density (lbm/bbl – ppb)
Barite Pure grade	4.5	37.5	
Barite API – drilling rade	4.2	35	1,470
Bentonite	2.6	21.7	910
Calcium carbonate	2.7	22.5	945
Calcium chloride	1 .96	16.3	686
Sodium chloride	2.16	18	756
Water	J	8.33	1,001
Diesel	0.86	7.2	300
Iron oxide	5	41.73	5,005
Galena	6.6	55	6,007
Drilled solids	2.7	21.7	2,702

Table 4.3 Micron sizes.

Mineral	Micron size
Clay	<1
Bentonite	<1
Barite	2–60
Silt	2.74
Sand	>74

Table 4.4 Micron Cut Points for Solid Removal Systems.

Equipment	**Micron size**
Centrifuge	3 to 5
Desilter (3 to 4-inch cones	12 to 60
Desander (5 to 12-inch cones)	30 to 60

Example calculation
Determine the percentage of clay based on a ton (2000 lb) of clay of 10 ppg clay-water drilling mud. The final mud volume is 100 barrels.

Solution:
If *X* is the percentage of clay in the final mud, then:

$$X/100 \times \rho_f \times V_f = 2000$$

$$X/100 \times 10 \times 500 = 2000$$

Therefore:

$$X = 2000/(10 \times 100 \times 42) \times 100 = 4.76\%$$

4.04 Diluting Mud

The dilution process involves selective dumping of the active system (such as sand traps and bottoms-up mud) and replacement of the lost volume with fresh mud. This process has proved economical with inhibitive water-base systems and is the only method that actually removes colloidal-size particles.

When diluting mud with a liquid, the resulting density of the diluted mud can be derived from:

$$\rho_f = \rho_o + V_a/V_o(\rho_o - \rho_a)$$

ρ_o is the original mud weight, ppg, V_a is the original volume in barrels, V_o is the final volume in barrels, and ρ_a is the density of material added, ppg.

Example calculation
Determine the final mud density of mud if 100 barrels of diesel with a specific gravity of 0.82 are mixed with 500 barrels 9 ppg mud.

Solution:
The final volume of mud after adding diesel = 500 + 100 = 600 bbl. Using the density relationship, the final density of the mud can be derived from:

$$\rho_f = [9 + (100/6)](9 - 0.82 \times 8.33) = 8.57 \text{ ppg}$$

4.05 Drilling Fluid Composition

A drilling fluid (also called drilling mud) is a fluid that is used to aid the drilling of of crude oil and natural gas. The three main categories of drilling fluids are water-based muds (which can be dispersed and non-dispersed), non-aqueous muds, usually called oil-based mud, and gaseous drilling fluid, in which a wide range of gases can be used.

Element	Water	Cuttings	Barite	Clay	Chrome-ligno-sulfonate	Lignite	Caustic
Aluminum	0.3	40,400	40,400	88,600	6,700	6,700	0.013
Arsenic	0.0005	3.9	34	3.9	10.1	10.1	0.039
Barium	0.01	158	590.000	640	230	230	0.26
Calcium	15	240,000	7,900	4,700	16,100	16.100	5,400
Cadmium	0.0001	0.08	6	0.5	0.2	0.2	0.0013
Chromium	0.001	183	183	8.02	40.030	65.3	0.00066
Cobalt	0.0002	2.9	3.8	2.9	5	5	0.00053
Copper	0.003	22	49	8.18	22.9	22.9	0.039
Iron	0.5	21.900	12,950	37,500	7.220	7.220	0.04
Lead	0.003	37	685	27.1	5.4	5.4	0.004
Magnesium	4	23,300	3,900	69,800	5,040	5,040	17.800
Mercury	0.0001	0.12	4.1	0.12	0.2	0.2	5
Nickel	0.0005	15	3	15	11.6	11.6	0.09
Potassium	2.2	13,500	660	2,400	3.000	460	51.400
Silicon	7	206,000	70,200	271.000	2,390	2,390	339
Sodium	6	3,040	3.040	11,000	71.000	2,400	500.000
Strontium	0.07	312	540	60.5	1030	1030	105

4.06 Equivalent Mud Weight

On a theoretical basis, the equivalent mud weight and the hydrostatic pressure are the same if the borehole is free of cuttings. Typically, the borehole is not free of cuttings and the annular pressure sensor is used to provide an accurate downhole pressure reading used to calculate equivalent mud weight (EMW). However, the true vertical depth (TVD) must be correct to effectively use the equivalent mud weight to make real-time drilling decisions.

Thus, the pressure in the wellbore or formation can be expressed in terms of equivalent mud weight (EMW), which is a convenient way to compare the pressures at any depth.

$$\text{EMW (ppg)} = P_h/(0.52 \times L_{tvd})$$

L_{tvd} is the true vertical depth (TVD) in feet, and P_h is the annular pressure in psi. If the well is deviated α degrees from the vertical, then:

$$\text{EMW (ppg)} = P_h/(0.52 \times D_h \cos \alpha)$$

D_h is the measured depth in feet.

4.07 Fluid Loss

The volume fraction of solids in mud, f_{vm}, can be written as

$$f_{vm} = V_s/V_m = f_{vc} + hA/(V_f + hA)$$

V_f is the amount of filtrate volume, h is the cake thickness, A is the filtration area, V_m is the total volume of mud filtered, and V_s is the volume of solids deposited in mud cake.

The API water loss is derived from:

$$V_{30} = V_{7.5} - V_{sp}$$

Filter loss is always estimated with reference to the square root of time. Filtrate volume is derived from:

$$V_f = A[2k(f_{vc}/f_{vm} - 1)\Delta Pt/\mu]^{1/2} + V_s$$

where ΔP is the differential pressure, μ is the filtrate viscosity, t is the filtration time, and k is the cake permeability.

Example calculation

The spurt loss of a mud is known to be 2 cc. If a filtrate of 10 cc is collected in 15 minutes using a filter press, what is the standard API filtrate for this mud?

Solution:

The filter loss at 7.5 minutes is found by extrapolating the data given at time zero and fifteen minutes. Thus, if X is the filter loss at 7.5 minutes:

$$2 = (10 - X)(15^{1/2} - 7/5^{1/2}) \times 7.5^{1/2}$$

Thus:

$$V_{30} = 2V_{7.5} - V_{sp} = 2 \times 6.69 - 10 = 3.38 \text{ cc}$$

Let x be the filter loss at 7.5 minutes. So, the filter loss at time 7.5 minutes is

$$2 = \frac{10 - x}{\sqrt{15} - \sqrt{7.5}} \times \sqrt{7.5}$$

Solving for × will yield 6.69 cc.
Using equation 4.29, API filter loss is

$$V_{30} = 2V_{7.5} - V_{sp} = 2 \times 6.69 - 10 = 3.38cc.$$

Problem 4.32

A kick was taken at 10000' while drilling with a 14 ppg mud. Stabilized shut-in drill pipe pressure = 600 psi. Determine the amount of barite to be added to 800 bbl of original mud in order to contain formation pressure such that a differential pressure of 300 psi is achieved after killing the well.

Solution:

Equivalent mud weight needed to control the kick with the excess differential pressure is

$$\frac{600 + 0.052 \times 14 \times 10000(7280) + 300}{0.052 \times 10000} = 15.73.$$

The amount of barite required is

$$m_g = 42 \times 28.02 \frac{(15.73 - 14)}{(28.02 - 15.73)} = 132593 \text{ lbm}.$$

Assuming 100 lbm per sack, the number of sacks of barite required is 1325.

4.08 Marsh Funnel

The Marsh funnel is a device for measuring viscosity by observing the time it takes a known volume of liquid to flow from a cone through a short tube. It is standardized for use by mud engineers to check the quality of drilling mud. As a standard reference point, the time for fresh water to drain = 26 sec ± 0.5 sec per quart for API water at 70°F ± 0.5°F.

The effective viscosity can be determined from following simple formula:

$$\mu = \rho \, (t - 25)$$

where μ is the effective viscosity in centipoise, ρ is the density in g/cm^3, t is the quart funnel time in seconds.

For example, a mud of funnel time 40 seconds and density 1.1 g/cm^3 has an effective viscosity of about 16.5 cP. For the range of times of typical muds above, the shear rate in the Marsh funnel is about 2000^{-s}.

4.09 Mud Rheology

Rheological properties of drilling fluid have considerable impact on carrying out of drilling cutting on a day surface, creation of hydrodynamic pressure in well. In terms of mud rheology, the shear stress (τ) and shear rate (γ) for Newtonian fluid are given as:

$$\tau = \mu\gamma$$

μ is the Newtonian viscosity.

In engineering units, τ is dynes/cm^2 = 4.79 lb/100 feet^2, γ = sec^{-1}, and μ = poise = dyne × sec/cm^2. The field unit of viscosity is the centipoise (1 poise = 100 centipoise) and the field unit of shear stress is lb/100 ft^2.

For the Bingham plastic model, the relationship is given by the equation:

$$\tau = \tau_y + \mu_p \gamma$$

For the power law model, shear stress (τ) and shear rate (γ) are given by the equation:

$$\tau = K\gamma^n$$

Shear rate and shear stress relationship for various fluids are presented in Figures 4.1, 4.2, and 4.3.

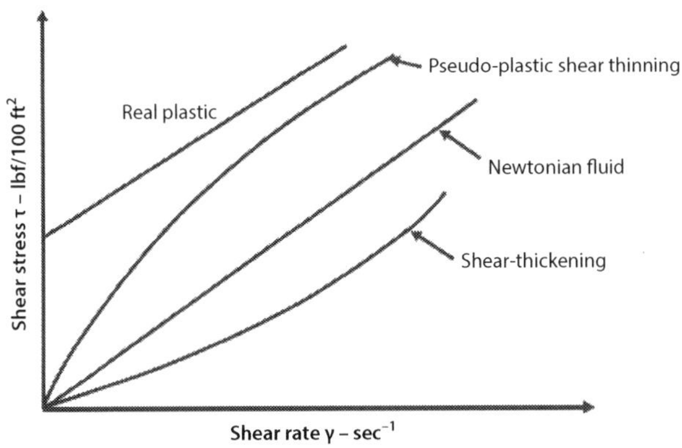

Figure 4.1 Shear Rate-Shear Stress Relationship of Time Independent Non-Newtonian Fluids.

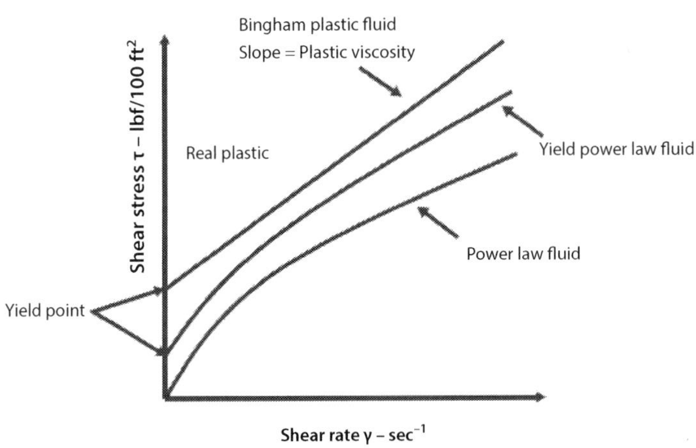

Figure 4.2 Shear Rate-Shear Stress Relationship of Non-Newtonian Fluids.

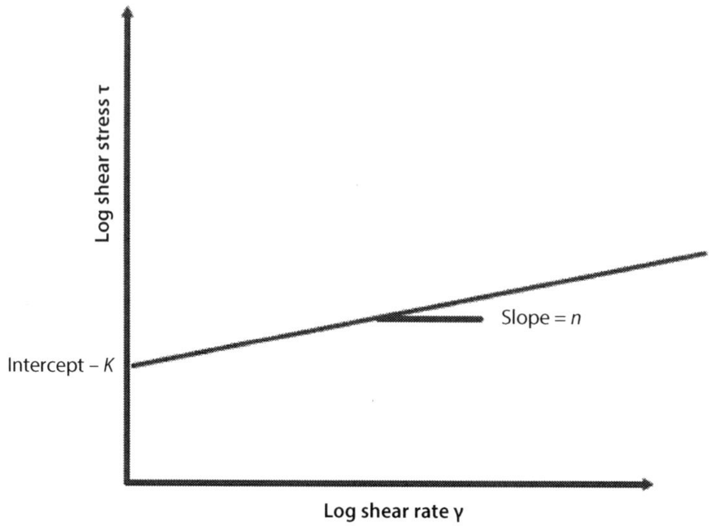

Figure 4.3 Shear Rate-Shear Stress Relationship of a Power Law Fluid Using the Logarithmic Scale.

For the yield power law (Herschel Bulkley), shear stress (τ) and shear rate (γ) are given by the equation:

$$\tau = \tau_y + K\gamma_n$$

τ_y is the yield value or yield stress, μ_p is the Bingham plastic viscosity, K is the consistency index, and n is the power law index.

4.10 Mud Weighting

Mud weighting is the addition of a high-density material such as barite to increase the mud weight (mud density)

The four fundamental equations used in developing mud weighting mathematical relations are (1) the material balance equation, (2) the volume balance equation, (3) the relationship between weight and volume equation, and (4) the volume balance in low specific gravity solids equation:

$$W_f = W_o + W_a$$

$W_f = W_o + W_a$ are the final mixture weight, original liquid weight, and added material weight, respectively. Thus:

$$V_f = V_o + V_a$$

V_f, V_o, and V_a are the final mixture volume, original liquid volume, and added material volume, respectively.

$$\rho = W/V$$

where ρ, W, and V are the weight density, weight, and volume, respectively.

$$V_f f_{vf} = V_o f_{vo}$$

where f_{vf} and f_{vo} are the volume fraction of low specific gravity solids in the final mixture and in the original liquid, respectively.

The above equations may be combined and rearranged algebraically to suit the calculations desired. Thus:

$$V_1\rho_1 + V_2\rho_2 + V_3\rho_3 + V_4\rho_4 + \ldots = V_f\rho_f$$

$$V_1 + V_2 + V_3 + V_4 + \ldots = V_f$$

$V_1 + V_2 + V_3 + V_4$ are the volumes of materials 1, 2, 3, and 4, respectively, and ρ_1, ρ_2, ρ_3 and ρ_4 are the densities of materials 1, 2, 3, and 4, respectively.

A simultaneous solution of the two equations above results in any two sought unknowns and the rest of the parameters are known. The following are example formulations for commonly encountered field problems:

1. The amount of weighting materials required to increase original mud density, ρ_o, to a final density, ρ_f, is

   $$W_{wm} = 42(\rho_f - \rho_o)/(1 - \rho_f/\rho_{wm})$$

   W_{wm} is the required amount of weighting material, lbs/bbl, or original mud.
2. The average weight density, ρ_{av}, of two added materials i and j of weights ρ_i and ρ_j, respectively, is:

   $$\rho_{av} = \rho_i\rho_j/f_w f_i + (1 + f_w)\ \rho_j$$

where $f_w = w_j/w_i + w_j$ is the weight fraction of material j with respect to added weights of materials i and j.

3. The amount of liquid volume, V_l, required to make a mud of total volume V_f having a weight of ρ_f is:

$$V_l = V_f[(1 - \rho_f/\rho_a)/(1 - \rho_l/\rho_a)]$$

where ρ_a is the weight density of added material or average weight density, ρ_{av}, of material added to the liquid.

4. The amount of solids i and j, such as clay and barite for example, required to make-up a specified final mud volume, V_f, and density, ρ_f, is:

$$wi = 42f_w(\rho_f - \rho_o)/1 - (\rho_o/\rho_j) - (1 - f_w)(\rho_o\rho_i)$$

$$wj = (1 - f_w)(\rho_o - \rho_j)/1 - (\rho_o/\rho_j) - (1 - f_w)(\rho_o\rho_i)$$

where w_i and w_j are the solid materials i and j in lbs/bbl of final mud.

5. The final mud density, ρ_f, when a certain liquid volume, V_l, of density ρ_l is added to a mud system of original density, ρ_o, and volume, V_o, is:

$$\rho_f = (\rho_o + \alpha\rho_l)/(1 + \alpha)$$

where $\alpha = V_l/V_o$.

4.11 Plastic Viscosity, Yield Point, and Zero-Sec Gel

4.11.1 Bingham Plastic Model

Plastic viscosity, yield point, and zero-sec-gel can be calculated from the Fann reading using the following relationships:

$$PV = \Theta_{600} - \Theta_{300}$$

$$YP = 2\Theta_{300} - \Theta_{600}$$

$$\tau = \Theta_3$$

Alternatively, the dial readings can be reverse calculated by using PV, YP, and zero-gel. Thus:

$$\Theta_{300} = PV + YP$$

$$\Theta_{600} = 2PV + YP$$

$$\Theta_{3} = \tau_{0}$$

Θ_{600} is the Fann dial reading at 600 rpm, Θ_{300} is the Fann dial reading at 300 rpm, and Θ_3 is the dial reading at 3 rpm. If the Fann RPMs are anything other than 600 rpm and 300 rpm, the following relationship can be used:

$$\mu_p = 300/(N_2 - N_1)/(\Theta_{N1} - \Theta_{N2})$$

4.11.2 Shear Stress and Shear Rate

Shear stress and shear rate can be calculated using the following relationships:

$$\tau_1 = (0.01065)\ \Theta\ \text{lbf/ft}^2$$

$$\tau_2 = (1.065)\ \Theta\ \text{lbf/100 ft}^2$$

$$\gamma = (1.073)\ N\ 1/\text{sec}$$

N is the dial speed in 1/sec.

4.11.3 Power Law

The rheological equation for the power law model can be given as:

$$\tau = K\gamma^n$$

where γ is the shear rate (1/sec) and τ is the = the shear stress ($1b/ft^2$).

The flow behavior index can be given as:

$$n = 3.322 \log (\Theta_{600}/\Theta_{300})$$

For the modified power law, the equation is:

$$n = 3.322\log (YP + 2PV)/(YP + PV)$$

PV is the plastic viscosity and YP is the yield point.

The consistency index, K, is given as:

$$K = (510\Theta_{300})/511n \text{ eq. cP}$$

K can be expressed in ($lb \times sec^n/ft^2$) using the conversion factor:
The consistency index, K, for the modified power law is given as:

$$K = (YP + 2PV)/[(100)(1022^n)$$

If the Fann RPMs are anything other than 600 and 300, the following relationship can be used:

$$n = (\log \Theta_{N2}/\Theta_{N1})/(\log N_2/N_1)$$

4.12 Reynolds Number and Critical Velocity

In fluid mechanics, the Reynolds number (Re) is a dimensionless quantity that is used to help predict flow patterns in different fluid flow situations. The Reynolds number is used to predict the transition from laminar flow to turbulent flow.

The Reynolds Number, used in the annular Power Law Model calculations is calculated using equivalent viscosity (m):

$$M = [90000 \times Pl_a \times (d_1 - d_2)^2]/(L \times V)$$

The Reynolds Number is then:

$$Re = [15.47 \times MD \times V \times (d_1 - d_2)]/\mu$$

The fluid velocity that will produce the critical Reynolds Number for given fluid properties and pipe configuration is found using:

$$Vc = 60 \times [(Re_L \times k)/(185.6 \times MD)] \times [(96 \times G)/(d_1 - d_2)^n]^{1/(2-n)}$$

where: Re_L = Laminar/Transitional Reynolds Number.

4.13 Slip Velocity

The slip velocity is the difference between the average velocities of two different fluids flowing together in a pipe. In vertical ascending flow, the lighter fluid flows faster than the heavier fluid. Thus, the slip velocity depends mainly on the difference in density between the two fluids, and their relative holdup. The slip velocity of a cutting in turbulent flow may be estimated using the following equation:

$$Vs = 113.4 \times [d_p(pp - MD)/(CD \times MD)]^{0.5}$$

where: Vs = Slip Velocity (ft/min)
d_p = Particle Diameter (inches)
pp = Particle density (lb/gal)
MD = Mud Density (lb/gal)
CD = Drag Coefficient

For these calculations, the particle density is calculated by multiplying the cuttings density (gm/cc) by the density of fresh water. The drag coefficient is the frictional drag between the fluid and the particle. In turbulent flow, the drag coefficient is 1.5. In laminar flow, the equivalent viscosity (m) will influence the slip velocity. In this case the slip velocity is:

$$V_s = 175.2 \times d_p \times [pp - MD)^2/(\mu \times MD)]^{0.333}$$

5

Hydraulics

5.01 Basic Calculations

5.01.1 Critical Velocity

The critical velocity is the velocity at which the flow regime changes from laminar flow to turbulent flow. For the Bingham plastic model:

$$V_c = [1.08PV + 1.08(PV^2 + 12.34\rho_m D_i^2 YP)^{1/2}]\rho D_i$$

$$V_c = [1.08\mu_p + 1.08(\mu_p^2 + 12.34\rho_m D_i^2 \tau_y)^{1/2}]\rho D_i$$

PV (or μ_p) is the plastic viscosity in centipoise, YP (or τ_y) is the yield point lbf/100 ft^2, and ρ_m is the mud density in ppg.

5.01.2 Pump Calculations

Pump pressure is obtained from the equation:

$$P_p = \Delta p_b + P_d \text{ psi}$$

Δp_b is the bit pressure drop in psi and P_d is the loss die to frictional pressure, also in psi. The hydraulic horsepower (HHP) of the bit is:

$$(Q \times \Delta p_b)/1714 \text{ hp}$$

Q is the flow rate in gallons per minute and Δp_b is the decrease in bit pressure in psi.

5.02 Bingham Plastic Model

The Bingham plastic model uses the Reynolds number (R_e) which is a dimensionless quantity that is used to help predict flow patterns in different fluid flow situations, especially the transition from laminar flow to turbulent flow as well as for the scaling of similar but different-sized flow situations. The predictions of the onset of turbulence and the ability to calculate scaling effects can be used to help predict fluid behavior on a larger scale, such as in local or global air or water movement and thereby the associated meteorological and climatological effects.

For flow in a pipe, the Reynolds number is calculated using the relationship:

$$N_{Rep} = (928\ \rho_m\ v_p D_i)/\mu_{ep}$$

where ρ_m is the mud weight in ppg, v_p is the velocity of the fluid in fps.

$$v_p = Q/(2.448\ D_i^2)$$

Q is the glow rate in gallons per minute (gpm), D_i is the inside diameter of the pipe, and μ_{ep} is the equivalent viscosity of the fluid. Thus:

$$\mu_{ep} = \mu_p + (20\ D_i\ \tau_y)/3\ v_p$$

μ_p is the plastic viscosity in centipoise and τ_y is the yield point lbf/100 ft^2.

The pressure gradient for the unit length *dL* and for the laminar flow is calculated from the equation:

$$dp_f/dL = (\mu_p v_p)/(1500\ D_i^2) + \tau_y/(225\ D_i) \text{ psi/ft}$$

For flow in the annulus:

$$N_{Rea} = [(757\ \rho_m v_a)(D_2 - D_p)]/\mu_{ep}$$

$$\mu_{ep} = \mu_p + 5(D_2 - D_p)\tau_y/v_a$$

where D_2 is the annulus diameter in inches, D_p is the outside diameter of the pipe in inches, and v_a is the velocity of the fluid in fps:

$$v_a = Q/[2.448(D_2^2 - D_p^2)\ \text{ft/sec}$$

Thus, the pressure gradient for laminar flow is calculated from:

$$(dp_f/dL) = (\mu_p v_a)/[1000(D_2 - D_p)^2 + \tau y/[200(D_2 - D_p)]\ \text{psi/ft}$$

Example calculation
Calculate the change in the bottomhole pressure when the yield value is 5 lb/100 ft^2 and the viscosity is 30 cp. Use the following data:

- Hole size = 97/8 inches
- Depth = 10,000 ft
- Pipe OD = 4 inches
- Flow rate = 400 gpm
- Mud weight = 12 ppg
- Yield point = 60 lb/100 ft^2
- Plastic viscosity = 40 cp

Solution:
The annular velocity of the fluid is:

$$v_a = 400/[2.448(9.875^2 - 4.5^2)] = 2.11\ \text{ft/sec}$$

The equivalent viscosity is:

$$\mu_{ep} = 40 + [5 \times 60(9.875 - 4.50]/2.11 = 802.5\ \text{cP}$$

The Reynolds number is:

$$N_{Rea} = [7.57 \times 12 \times 2.11(9.875 - 4.5)]/802.5 = 128$$

Since the Reynolds's number is less than 2100, the flow is laminar.

The equivalent viscosity with new YP and PV is:

$\mu_{ep} = 5 + [5 \times 30(9.875 - 4.5)]/2.11 = 93.5$ cP and the Reynolds number (1104) is still indicating laminar flow. Thus:

$$dp_f/dL = [(40 - 5)v_a]/[1000(9.875 - 4.5)_2] + (60 - 30)/[200(9.875 - 4.5)] = 0.051985 \text{ psi/ft}$$

Thus, the total pressure change is $0.051985 \times 10{,}000 = 519$ psi

5.03 Bit Hydraulics

5.03.1 Common Calculations

The total nozzle flow is calculated from the equation:

$$A_n = (\pi \times S_n^2)/64 \text{ square inches}$$

The nozzle velocity is:

$$V_n = 0.3208(Q/A_n) \text{ ft/sec}$$

The pressure drop across bit is calculated from:

$$\Delta p_b = (8.311 \times 10^{-5}\, \rho_m\, Q^2)/(C_d^2 \times A_n^2) \text{ psi}$$

C_d is the discharge coefficient (usually a value of 0.95 is used). If the value 0.95 is used, the pressure drop across the bit is:

$$\Delta p_b = (\rho_m \times Q^2)/(10858 \times A_n^2) \text{ psi}$$

Using nozzle velocity, the pressure drop across the bit is calculated from:

$$\Delta p_b = (\rho_m \times V_n^2)/1120 \text{ psi}$$

The percentage pressure drop across the bit is:

$$\Delta p_b/p_b \times 100$$

The bit hydraulic power (BHHP) is calculated from:

$$HP_b = (Q \times \Delta p_b)/1714$$

Hydraulic bit (jet) impact force (F_{imp}) is calculated from:

$$F_{imp} = 0.01823 \times C_d \times Q(\rho_m \times \Delta p_b)^{1/2} \text{ lbf}$$

The hydraulic bit (jet) impact force (F_{imp}) can also be written as:

$$F_{imp} = [Q \times (\rho_m \times \Delta p_b)^{1/2}]/57.66 \text{ lbf}$$

The hydraulic bit (jet) impact force (F_{imp}) with nozzle velocity can be written as:

$$F_{imp} = (\rho_m \times Q \times V_m)/1930 \text{ lbf}$$

The impact force per square inch of the bit area is calculated from:

$$F_{imp}/[(\pi \times D_b^2)/4] \text{ lbf in}^2$$

The impact force per square inch of the hole area is:

$$F_{imp}/[(\pi \times D_h^2)/4] \text{ lbf in}^2$$

5.03.2 Optimization Calculations

The flow index is:

$$m = [\log (\Delta p_d)_j/(\Delta p_d)_i]/9\log Q_j/Q_i)$$

For maximum bit hydraulic power:

$$\Delta p_{bopt} = [m/(m + 1)] \times \Delta p_{max}$$

The optimum flow rate is calculated from:

$$Q_{opt} = Q_a \text{alog}[1/m \log(\Delta p_{dopt}/\Delta p_{dQa})$$

5.03.2.1 Limitation 1 – Available Pump Horsepower

For maximum impact force, the optimum bit pressure drop is:

$$\Delta p_{bopt} = [(m + 1)/(m + 2)]\ \Delta p_{bopt}$$

The optimum flow rate is:

$$Q_{opt} = [92 \times \Delta p_{max})/c(m + 2)]^{1/m}$$

5.03.2.2 Limitation 2 – Surface Operating Pressure

For maximum impact force, the optimum bit pressure drop is:

$$\Delta p_{bopt} = m/(m + 2)]\Delta p_{max}$$

Example calculation
Estimate the optimum nozzle size and optimum flow rate for the following conditions:

- *m* = 1.66
- Maximum allowed operating pressure = 5440 psi
- Frictional pressure loss = 2334 psi for a flow rate of 300 gpm
- Volumetric efficiency of the pump = 80%
- Mud weight = 15.5 ppg
- Minimum flow rate required for hole cleaning = 265 gpm

Solution:
Using limitation 2 (above):

$$\Delta p_{bopt} = m/(m + 2)]\Delta p_{max}$$

$$\Delta p_{bopt} = 2/(166 + 2) \times 5440 = 2992 \text{ psi}$$

$$\Delta p_{bopt} = \Delta p_{max} - \Delta p_{dopt}$$

$$\Delta p_{bopt} = 5440 - 2975 = 2465 \text{ psi}$$

5.04 Critical Transport Fluid Velocity

The critical transport fluid velocity (CTFV) is the minimum fluid flow velocity in a pipe or annular region required to prevent the formation of a stationary cuttings bed. The CTFV is calculated using the equation:

$$V_{ca} = V_{cr} + V_{cs}$$

V_{ca} is the critical transport average annular fluid velocity, V_{cr} is the cuttings rise velocity, and V_{cs} is the cuttings average slip velocity.

5.05 Equivalent Circulating Density

The equivalent circulating density (ECD) results from the addition of the equivalent mud weight, due to the annulus pressure losses (Δp_a), to the original mud weight (ρ_m). Thus:

$$ECD = \rho_m + \Delta p_a/(0.052 \times L_{tvd}) \text{ ppg}$$

Example calculation
Calculate the equivalent mud weight at a depth of 10,000 ft (TVD) with an annulus back pressure of 500 psi. The mud density in the annulus is 10 ppg.

Solution

$$EMW = \rho_m + 500/(0.052 \times 10{,}000) = 10.96 \text{ ppg}$$

5.06 Equivalent Mud Weight

The pressure in the wellbore or formation can be expressed in terms of equivalent mud weight (EMW) whohc is a convenient way in which to compare the pressure at any depth. This:

$$EMW = P_h/(0.052 \times L_{tvd}) \text{ ppg}$$

Ph is the pressure in psi, and Ltvd is the true vertical depth in feet. If the well is deviated, then:

$$EMW = P_h/(0.052 \times D_h \cos \alpha) \text{ ppg}$$

D_h is the measured depth in feet and α is the deviation in degrees from the vertical.

5.07 Gel Breaking Pressure

In the pipe, the gel breaking pressure (P) is calculated from the equation:

$$P = (\tau_g L)/(300\ D_i)\ \text{psi}$$

D_i is the inside diameter of the pipe, τ_g is the gel strength (lbf/100 ft^2) and L is the length of the pipe in feet.

In the annulus:

$$P = (\tau_g L)/300(D_h - D_p)\ \text{psi}$$

D_h is the diameter of the hole in inches and D_p is the outside diameter of the pipe, also in inches.

Example calculation

Mud circulation has been stopped for sufficient time to develop a gel strength of 15 lbf/100ft^2. If the pipe is not moved, calculate the pressure surge required to break the circulation.

- Depth = 10,000 ft
- Drill pipe = 5 inches × 4.27 inches
- Drill collar = 450 feet, 6.5 inches × 3 inches
- Hole diameter = 8.5 inches

Solution:

The pressure (P) required to break circulation is

$$P = [(15 \times 9550)/(300 \times 4.27)] + [(15 \times 450)/(300 \times 3)] + [(15 \times 9550)/300(8.5 - 5)] + [(15 \times 450)/300(8.5 - 6.5)]$$

$$P = 1181.8 + 7.5 + 136.43 + 11.25 = 267\ \text{psi}$$

5.08 Hole Cleaning – Cuttings Transport

Drill cuttings are the broken bits of solid material removed from a borehole. The cuttings are commonly examined to make a record (a well log or sometimes called a mud log) of the subsurface materials penetrated at various depths. The cuttings concentration for 45° with one tool is given by the equation:

$$C_c = 3.22(1 + N_{Ta})^{-0.472} + 5703.6N_{Re}^{-0.776} + 69.3_{Rop}^{-0.051}$$

$$(1450 < N_{Re} < 3700, 0 < N_{Ta} < 5800, 19.7 < N_{Rop} < 23)$$

The cuttings concentration for 90° with one tool is given by the equation:

$$C_c = -5.22 \times 10^{-5}N_{Ta}^{1.36} + 605.714N_{Re}^{-0.0124} + 8.86e^{230.43}N_{Rop}^{-77.58} - 529.4$$

$$(1450 < N_{Re} < 3700, 0 < N_{Ta} < 5800, 19.7 < N_{Rop} < 23)$$

For the above case with no tools and with a 45° inclination:

$$C_c = 1.81(1 + N_{Ta})^{-8.13} + 21.25 \times e^{177.57}N_{Re}^{-24.63} + 15.7 \times e^{3.54}N_{Rop}^{-0.03} - 487.57$$

$$(1450 < N_{Re} < 3700, 0 < N_{Ta} < 5800, 19.7 < N_{Rop} < 23)$$

The cuttings concentration for 90° with no tool is given by the equation:

$$C_c = -14.04 \times e^{-316}(1 + N_{Ta})^{36.33} + 284.96N_{Re}^{-0.073} + 58.2 \times e^{30.9}N_{Rop}^{-12.46} - 140.88$$

$$(1450 < N_{Re} < 3700, 0 < N_{Ta} < 5800, 19.7 < N_{Rop} < 23)$$

In the above equations, N_{Ta} is Taylor's number:

$$N_{Ta} = \omega\rho D^2/\mu_{eff}$$

D is the diameter of the hole in inches, ω is the angular velocity of the tool in radians per second, ρ_m is the mud density in ppg, and μ_{eff} is the effective viscosity in centipoise.

N_{Re} is the Reynold number $= (\rho V^{2-n}D^n)/K8^{n-1}$

K is the consistency index, V is the velocity of the fluid, and n is the power law index.

N_{Rop} is the ROP number (rate of penetration): $N_{Rop} = (\rho D \times ROP)/\mu_{eff}$

6

Tubular Mechanics

6.01 API Casing and Liners – Weight, Dimensions, Capacity, and Displacement

Nominal size inches	Weight with coupling lb/ft	Outside diameter inches	Inside diameter inches	Capacity bbl/ft	Capacity ft/bbl	Displacement bbl/100 ft
4-1/2	9.50	4.500	4.090	0.0162	61.54	0.342
4-1/2	10.50	4.500	4.052	0.0159	62.70	0.372
4-1/2	11.60	4.500	4.000	0.0155	64.34	0.413
4-1/2	13.50	4.500	3.920	0.0149	67.00	0.474
4-1/2	15.10	4.500	3.826	0.0142	70.33	0.545
5	11.50	5.000	4.560	0.0202	49.51	0.409
5	13.00	5.000	4.494	0.0196	50.97	0.467
5	15.00	5.000	4.408	0.0189	52.98	0.541
5	18.00	5.000	4.276	0.0178	56.30	0.652

(Continued)

Nominal size inches	Weight with coupling lb/ft	Outside diameter inches	Inside diameter inches	Capacity bbl/ft	Capacity ft/bbl	Displacement bbl/100 ft
5	21.40	5.000	4.126	0.0165	60.47	0.775
5	23.20	5.000	4.044	0.0159	62.95	0.840
5	24.10	5.000	4.000	0.0155	64.34	0.874
5-1/2	14.00	5.500	5.012	0.0244	40.98	0.498
5-1/2	15.50	5.500	4.950	0.0238	42.02	0.558
5-1/2	17.00	5.500	4.892	0.0232	43.02	0.614
5-1/2	20.00	5.500	4.778	0.0222	45.09	0.721
5-1/2	23.00	5.500	4.670	0.0212	47.20	0.820
5-1/2	26.80	5.500	4.500	0.0197	50.84	0.971
5-1/2	29.70	5.500	4.376	0.0186	53.76	1.078
5-1/2	32.60	5.500	4.250	0.0175	57.00	1.184
5-1/2	35.30	5.500	4.126	0.0165	60.47	1.285
5-1/2	38.00	5.500	4.000	0.0155	64.34	1.384
5-1/2	40.50	5.500	3.876	0.0146	68.53	1.479
5-1/2	43.10	5.500	3.750	0.0137	73.21	1.572
6-5/8	20.00	6.625	6.049	0.0355	28.14	0.550
6-5/8	24.00	6.625	5.921	0.0341	29.37	0.699
6-5/8	28.00	6.625	5.791	0.0326	30.70	0.846
6-5/8	32.00	6.625	5.675	0.0313	31.97	0.976
7	17.00	7.000	6.538	0.0415	24.08	0.608
7	20.00	7.000	6.456	0.0405	24.70	0.711
7	23.00	7.000	6.366	0.0394	25.40	0.823
7	26.00	7.000	6.276	0.0383	26.14	0.934
7	29.00	7.000	6.184	0.0371	26.92	1.045
7	32.00	7.000	6.094	0.0361	27.72	1.152
7	35.00	7.000	6.004	0.0350	28.56	1.258
7	38.00	7.000	5.920	0.0340	29.37	1.355
7	42.70	7.000	5.750	0.0321	31.14	1.548
7	46.40	7.000	5.625	0.0307	32.54	1.686
7	50.10	7.000	5.500	0.0294	34.03	1.821
7	53.60	7.000	5.376	0.0281	35.62	1.952
7	57.10	7.000	5.250	0.0268	37.35	2.082
7-5/8	24.00	7.625	7.025	0.0479	20.86	0.854

(Continued)

Nominal size inches	Weight with coupling lb/ft	Outside diameter inches	Inside diameter inches	Capacity bbl/ft	Capacity ft/bbl	Displacement bbl/100 ft
7-5/8	26.40	7.625	6.969	0.0472	21.20	0.930
7-5/8	29.70	7.625	6.875	0.0459	21.78	1.056
7-5/8	33.70	7.625	6.765	0.0445	22.49	1.202
7-5/8	39.00	7.625	6.625	0.0426	23.46	1.384
7-5/8	42.80	7.625	6.501	0.0411	24.36	1.542
7-5/8	45.30	7.625	6.435	0.0402	24.86	1.625
7-5/8	47.10	7.625	6.375	0.0395	25.33	1.700
7-5/8	51.20	7.625	6.251	0.0380	26.35	1.852
7-5/8	55.30	7.625	6.125	0.0364	27.44	2.003
7-3/4	46.10	7.750	6.560	0.0418	23.92	1.467
8-5/8	24.00	8.625	8.097	0.0637	15.70	0.858
8-5/8	28.00	8.625	8.017	0.0624	16.02	0.983
8-5/8	32.00	8.625	7.921	0.0609	16.41	1.131
8-5/8	36.00	8.625	7.825	0.0595	16.81	1.278
8-5/8	40.00	8.625	7.725	0.0580	17.25	1.429
8-5/8	44.00	8.625	7.625	0.0565	17.71	1.578
8-5/8	49.00	8.625	7.511	0.0548	18.25	1.746
9-5/8	32.30	9.625	9.001	0.0787	12.71	1.129
9-5/8	36.00	9.625	8.921	0.0773	12.94	1.268
9-5/8	40.00	9.625	8.835	0.0758	13.19	1.417
9-5/8	43.50	9.625	8.755	0.0745	13.43	1.553
9-5/8	47.00	9.625	8.681	0.0732	13.66	1.679
9-5/8	53.50	9.625	8.535	0.0708	14.13	1.923
9-5/8	58.40	9.625	8.435	0.0691	14.47	2.088
9-5/8	59.40	9.625	8.407	0.0687	14.57	2.133
9-5/8	64.90	9.625	8.281	0.0666	15.01	2.338
9-5/8	70.30	9.625	8.157	0.0646	15.47	2.536
9-5/8	75.60	9.625	8.031	0.0626	15.96	2.734
10-3/4	32.75	10.750	10.192	0.1009	9.91	1.135
10-3/4	40.50	10.750	10.050	0.0981	10.19	1.414
10-3/4	45.50	10.750	9.950	0.0962	10.40	1.609
10-3/4	51.00	10.750	9.850	0.0942	10.61	1.801

(Continued)

Nominal size inches	Weight with coupling lb/ft	Outside diameter inches	Inside diameter inches	Capacity bbl/ft	Capacity ft/bbl	Displacement bbl/100 ft
10-3/4	55.50	10.750	9.760	0.0925	10.81	1.972
10-3/4	60.70	10.750	9.660	0.0906	11.03	2.161
10-3/4	65.70	10.750	9.560	0.0888	11.26	2.348
10-3/4	73.20	10.750	9.406	0.0859	11.64	2.631
10-3/4	79.20	10.750	9.282	0.0837	11.95	2.856
10-3/4	85.30	10.750	9.156	0.0814	12.28	3.082
11-3/4	42.00	11.750	11.084	0.1193	8.38	1.477
11-3/4	47.00	11.750	11.000	0.1175	8.51	1.657
11-3/4	54.00	11.750	10.880	0.1150	8.70	1.912
11-3/4	60.00	11.750	10.772	0.1127	8.87	2.140
11-3/4	65.00	11.750	10.682	0.1108	9.02	2.327
11-3/4	71.00	11.750	10.586	0.1089	9.19	2.525
13-3/8	48.00	13.375	12.715	0.1570	6.37	1.673
13-3/8	54.50	13.375	12.615	0.1546	6.47	1.919
13-3/8	61.00	13.375	12.515	0.1521	6.57	2.163
13-3/8	68.00	13.375	12.415	0.1497	6.68	2.405
13-3/8	72.00	13.375	12.347	0.1481	6.75	2.568
16	65.00	16.000	15.250	0.2259	4.43	2.277
16	75.00	16.000	15.124	0.2222	4.50	2.648
16	84.00	16.000	15.010	0.2188	4.57	2.982
16	109.00	16.000	14.688	0.2096	4.77	3.911
18-5/8	87.50	18.625	17.755	0.3062	3.27	3.074
20	94.00	20.000	19.124	0.3553	2.81	3.329
20	106.50	20.000	19.000	0.3507	2.85	3.788
20	133.00	20.000	18.730	0.3408	2.93	4.778

6.02 API Drill Pipe Capacity and Displacement

Nominal Size inches	Weight lb/ft	OD* inches	ID** Inches	Capacity bbl/ft	Capacity ft/bbl	Pipe Displacement bbl/100 ft
2-3/8	6.65	2.375	1.815	0.0032	311.43	0.228
2-7/8	10.40	2.875	2.151	0.0045	221.73	0.353

(Continued)

Nominal Size inches	Weight lb/ft	OD* inches	ID** Inches	Capacity bbl/ft	Capacity ft/bbl	Pipe Displacement bbl/100 ft
3-1/2	9.50	3.500	2.992	0.0087	114.60	0.320
3-1/2	13.30	3.500	2.764	0.0074	134.29	0.448
3-1/2	15.50	3.500	2.602	0.0066	151.53	0.532
4	11.85	4.000	3.476	0.0118	84.91	0.381
4	14.00	4.000	3.340	0.0109	91.96	0.471
4-1/2	13.75	4.500	3.958	0.0153	65.49	0.445
4-1/2	16.60	4.500	3.826	0.0143	70.08	0.545
4-1/2	20.00	4.500	3.640	0.0129	77.43	0.680
5	16.25	5.000	4.408	0.0189	52.80	0.541
5	19.50	5.000	4.276	0.0178	56.11	0.652
5	25.60	5.000	4.000	0.0156	64.12	0.874
5-1/2	21.90	5.500	4.778	0.0223	44.94	0.721
5-1/2	24.70	5.500	4.670	0.0213	47.04	0.820

*OD: outside diameter

**ID: Inside diameter

6.03 Bending Stress Ratio

The bending stress ration (BSR) is the ratio of the box section modulus to the pin section modulus. Thus:

$$BSR = [(D^4 - b^4)/D]/[R_t^4 - d^4)/R_t]$$

D is the connection or tool outside diameter in inches, b is the thread root diameter of box threads at the end of the pin in inches, R_t is the thread root diameter of pin threads 0.75 inches from the shoulder of the pin in inches, and d is the pin inside diameter in inches.

Typical accepted ranges of BSR are 2.25 to 2.75 for critical service application, 2.0 to 3.0 for normal service application, and 1.9 to 3.2 for limited service application.

6.04 Buckling Force

Buckling is instability in a structure that that leads to failure and occurs when a structure is subjected to compressive stress. Buckling is characterized by a sudden sideways deflection of a structural member which may occur even though the stresses that develop in the structure are well below those needed to cause failure of the material of which the structure is composed.

The buckling force Fb is defined by the equation:

$$F_b = -F_a + P_iA_i - P_oA_o$$

F_b is the buckling force, F_a is the axial force (tension positive), P_i is the internal pressure, A_i is equal to πri^2 where r_i is the inside radius of the tubing, P_o is the external pressure, and A_o is equal to $\pi r_o^{\ 2}$ where r_o is the outside radius of the tubing.

The Paslay buckling force, F_p, is defined by the equation:

$$F_p = [(4EI\omega_c)/r]^{1/2}$$

Example calculation

The sample calculation given here is for determining the buckling of 2-7/8 inch, 6.5 lbm/ft tubing inside a 7 inch, 32 lbm/ft casing. The tubing is submerged in 10 lbm/gal packer fluid with no other pressures applied. Assume that an applied buckling force of 30,000 lbf is applied at the end of the string in a well with a 60° deviation from the vertical. The effect of the packer fluid is to reduce the tubing weight per unit length through buoyancy:

$$w_e = w + A_i\rho_i - A_o\rho_o$$

where *we* is the effective weight per unit length of the tubing, Ai is the inside area of the tubing, ρi is the density of the fluid inside the tubing, Ao is the outside area of the tubing, and ρo is the density of the fluid outside the tubing. Thus:

$$we = 6.5 \text{ lbm/ft} + (4.68 \text{ in}^2)\ (0.052 \text{ psi/ft/lbm/gal}),$$

$$= (10.0 \text{ lbm/gal}) - (6.49 \text{ in}^2)\ (0.052 \text{ psi/ft/lbm/gal})\ (10.0 \text{ lbm/gal}),$$

$$= 5.56 \text{ lbm/ft} = 0.463 \text{ lbm/in.}$$

6.05 Drag Force

In fluid dynamics, drag (also called air resistance, fluid resistance, or fluid friction) is the force acting opposite to the relative motion of any object moving with respect to a surrounding fluid. This can exist between two fluid layers (or surfaces) or a fluid and a solid surface. Even though the ultimate cause of a drag is viscous friction, the turbulent drag is independent of viscosity. The drag force is proportional to the velocity for laminar flow and the squared velocity for turbulent flow.

Thus, drag force is given by the equation:

$$F_d = F_s \times \mu \times (V_{ts}/V_{rs})$$

V_{ts} is the trip speed in inches per second, V_{rs} is the resultant speed in inches per second, F_s is the side or normal force in lbs, and μ is the coefficient of friction.

6.06 Drill Collar Length

The drill collar is a thick-walled tubular piece machined from a solid bar of steel, usually plain carbon steel but sometimes of nonmagnetic nickel-copper alloy or other nonmagnetic premium alloys. The drill collar size is defined by the equation:

$$D_{dc} = (2D_{csg} - D_b)$$

D_{dc} is the diameter of the drill collar, D_{csg} is the diameter of the casing coupling, and D_b is the diameter of the drill bit. Thus, the length of the drill collar is given by:

$$L_{dc} = (WOB \times DF)/(w_{dc} \times BF \times \cos \alpha)$$

WOB is the weight on the drill bit in lbs, DF is the design factor, w_{dc} is the unit weight of the collar in lbf/ft, BF is the buoyancy factor, and α is the wellbore inclination in degrees.

Example calculation

An 8½-inch, 22° hole is planned to be drilled and cased with 7-inch, 38 ppf, P-110 – BTC casing. The mud weight to be used is 12 ppg. The weight on bit desired is 25 kips. Calculate the size and length of drill collar required.

Solution:

Coupling OD of the 7-inch" 38 ppf BTC is 7.656 inches. The diameter of the drill collar is:

$$D_{dc} = 2 \times 7.656 - 8.5 = 6.812 \text{ inches}$$

The closest available drill collar sizes are 6¾ inches and 6½ inches. Typically, the 6½-inch drill collar is selected so that annular pressure losses against the drill collars, and thereby the equivalent circulating density, (ECD) are reduced.

6.07 Fatigue Ratio

The fatigue ratio (FR_F, also called the endurance ratio) is the ratio of the fatigue limit or fatigue strength to the static tensile strength of a material. Thus:

$$FR_F = (\sigma_b + \sigma_{buck})/\sigma_{fl}$$

In this equation, σ_b is the bending stress in psi, σ_{buck} = is the buckling stress in psi, and σ_{fl} is the fatigue limit.

For tension, the fatigue limit can be written as:

$$\sigma_{fl} = \sigma_{el}(1 - F_e/F_y)$$

where σ_{el} is the fatigue endurance limit of the pipe in psi, F_y is the yield tension in lbf, and F_e is the effective tension in lbf. For compression, the fatigue limit is:

$$\sigma_{fl} = \sigma_{el}$$

The axial force on the pipe has a remarkable impact on the fatigue failure. The force required to generate the yield stress can be written as:

$$F_y = \sigma_{ymin} \times A_e$$

where σ_{ymin} is the minimum yield of the pipe in psi, and Ae is the effective cross-sectional area in square inches.

6.08 Length Change Calculations

Changing the mode of a well (producer, injector, shut-in, or treating) causes changes in temperature and pressure inside and outside the tubing. This, in turn, can create length and force changes in the tubing string that can potentially affect the packer and downhole tools. The thermal induced stretch is calculated from the following equation:

$$\Delta L_t = L a_t \Delta t$$

where a_t is the coefficient of thermal expansion which is the fractional increase in length per unit rise in temperature, with units of in/in/F (with values of 6.9×10^{-6} for steel, 10.3×10^{-6} for aluminum, and 4.9×10^{-6} for titanium). Δt is the average temperature change in degrees Fahrenheit. If there is a linear variation of temperature along the wellbore, then:

$$\Delta L_t = \alpha[\Delta t_o + (\Delta t/\Delta z)(L^2/2)]$$

where z is the measured depth and Δ L is the measured calculation interval.

Example calculation
Compute the elongation of 10,000 ft of 7-in casing due to temperature if the bottomhole temperature (BHT) is 290°F and the surface temperature is 70°F.

Solution

Δt = (final BHT + final surface temp)/2 – (initial BHT + initial surface temp)/2

$\Delta t = 110°F$

$\Delta L_t = 10{,}000 \times 6.9 \times 10^{-6} \times 110 = 7.59$ feet

6.09 Maximum Permissible Dogleg

Dogleg severity is a measure of the amount of change in the inclination, and/or azimuth of a borehole, usually expressed in degrees per 100 feet of course length. In the metric system, it is usually expressed in degrees per 30 meters or degrees per 10 meters of course length. The maximum permissible dogleg and the axial stress are given by the equation:

$$\sigma_t = F_{dls}/A \text{ psi}$$

A is the cross-sectional area of the drill pipe body in square inches and F_{dls} is the buoyed weight supported below the dogleg in lbs.

The maximum permissible bending stress for a drill pipe of grade E when the tensile stress is less than or equal to 67,000 psi is given by:

$$\sigma_b = 19{,}500 - (10/67)\sigma_t - (0.6/670^2)(\sigma_t - 33500)^2$$

The maximum permissible bending stress for grade S drill pipe when the tensile stress is less than or equal to 133,400 psi is:

$$\sigma_b = 20{,}000[1 - (\sigma_t/145{,}000) \text{ psi}$$

The maximum possible dogleg severity is:

$$c = (432{,}000\ \sigma_b)/\pi ED_p\ (\tan h\ KL)/KL$$

$$K = (T/EI)^{1/2}$$

E = Young's modulus in psi = 30×10^6 psi for steel and 10.5×10^6 psi for aluminum.

6.10 Pipe Wall Thickness and other Dimensions

6.11 Slip Crushing

Slip crushing refers to the maximum load that can be placed on the slips without damaging the specific tubular subjected to a tensile load. The calculation is dependent on the slip design, length of die contact to the pipe, the drill pipe thickness and yield strength. Thus, the maximum allowable load that can be supported by the slip is:

$$F_{max} = F_y/\{[(2D_p^{\,2}fA_p)/(D_p^{\,2} - D_p^{\,2})A_s] + [(2D_p^{\,2}fA_p)/(D_p^{\,2} - D_i^{\,2})A_s)^2]\}^{1/2}$$

F_y is the tensile strength of the pipe, lbf, A_s is the contact area between slips and pipe ($A_s = \pi\, D_p L_s$) in square inches, A_p is the cross sectional area of the pipe in square inches, D_p is the outside diameter of pipe, in, D_i is the inside diameter of pipe in inches, L_s is the length of slips, in, and f is the lateral load factor of slips = (1- μ tan α)/(μ + tan α), where μ is the coefficient of friction between slips and bushings, and α is the slip taper angle in degrees.

Table 6.1 Dimensions for Schedule 10, 40, 80 Pipes

		Schedule (10, 40, 80) wall thickness (inches.)**					
		Sch.10		**Sch.40**		**Sch.80**	
Pipe size*	**O.D. (in.)**	**Wall (in)**	**I.D. (in)**	**Wall (in)**	**I.D. (in)**	**Wall (in)**	**I.D. (in)**
1/8 inch	0.41 od	0.049 in	0.312 id	0.07 in	0.269 id		
Weight (lbs/ft.)	Steel	0.183 lbs/ft		0.247 lbs/ft			
	Stainless						
	Aluminum						
1/4 inch	0.54 od	0.065 in	0.410 id	0.090 in	0.364 id	0.119 in	0.302 id
Weight (lbs/ft.)	Steel	0.333 lbs/ft		0.429 lbs/ft		0.535 lbs/ft	
	Stainless						
	Aluminum			0.147 lbs/ft			
3/8 inch	0.675 od	0.065 in	0.545 id	0.091 in	0.493 id	0.126 in	0.423 id
Weight (lbs/ft.)	Steel	0.420 lbs/ft		0.570 lbs/ft		0.740 lbs/ft	
	Stainless						
	Aluminum			0.196 lbs/ft			
1/2 inch	0.840 od	0.083 in	0.674 id	0.109 in	0.622 id	0.147 in	0.546 id
Weight (lbs/ft.)	Steel	0.670 lbs/ft		0.850 lbs/ft		1.090 lbs/ft	
	Stainless						
	Aluminum			0.294 lbs/ft		0.384 lbs/ft	

(Continued)

Table 6.1 Cont.

		Schedule (10, 40, 80) wall thickness (inches.)**					
		Sch.10		Sch.40		Sch.80	
Pipe size*	**O.D. (in.)**	**Wall (in)**	**I.D. (in)**	**Wall (in)**	**I.D. (in)**	**Wall (in)**	**I.D. (in)**
3/4 inch	1.050 od	0.083 in	0.884 id	0.113 in	0.824 id	0.154 in	0.742 id
Weight (lbs/ft.)	Steel	0.86 lbs/ft		1.13 lbs/ft		1.48 lbs/ft	
	Stainless						
	Aluminum			0.391		0.520	
1 inch	1.315 od	0.109 in	1.097 id	0.133 in	1.049 id	0.179 in	0.957 id
Weight (lbs/ft.)	Steel	1.41 lbs/ft		1.68 lbs/ft		2.17 lbs/ft	
	Stainless						
	Aluminum	0.481 lbs/ft		0.581 lbs/ft		0.781 lbs/ft	
1-1/4 inch	1.66 od	0.109 in	1.442 id	0.140 in	1.380 id	0.191 in	1.278 id
Weight (lbs/ft.)	Steel	1.81 lbs/ft		2.27 lbs/ft		3.00 lbs/ft	
	Stainless						
	Aluminum	0.618 lbs/ft		0.785 lbs/ft		1.040 lbs/ft	
1-1/2 inch	1.90 od	0.109 in	1.682 id	0.145 in	1.610 id	0.200 in	1.500 id
Weight (lbs/ft.)	Steel	2.09 lbs/ft		2.72 lbs/ft		3.63 lbs/ft	
	Stainless						
	Aluminum	0.714 lbs/ft		0.939 lbs/ft		1.260 lbs/ft	

2 inch	2.375 od	0.109 in	2.157 id	0.154 in	2.067 id	0.218 in	1.939 id
Weight (lbs/ft.)	Steel	2.64 lbs/ft		3.66 lbs/ft		5.03 lbs/ft	
	Stainless						
	Aluminum	0.903 lbs/ft		1.260 lbs/ft		1.740 lbs/ft	
2-1/2 inch	2.875 od	0.120 in	2.635 id	0.203 in	2.469 id	0.276 in	2.323 id
Weight (lbs/ft.)	Steel	3.53 lbs/ft		5.80 lbs/ft		7.67 lbs/ft	
	Stainless						
	Aluminum	1.209 lbs/ft		2.000 lbs/ft			
3 inch	3.50 od	0.120 in	3.26 id	0.216 in	3.068 id	0.30 in	2.90 id
Weight (lbs/ft.)	Steel	4.34 lbs/ft		7.58 lbs/ft		10.26 lbs/ft	
	Stainless						
	Aluminum	1.483 lbs/ft		2.620 lbs/ft		3.55 lbs/ft	
3-1/2 inch	4.00 od	0.120 in	3.76 id	0.226 in	3.550 id	0.318 in	3.360 id
Weight (lbs/ft.)	Steel	4.98 lbs/ft		9.12 lbs/ft		12.52 lbs/ft	
	Stainless						
	Aluminum			3.150 lbs/ft		4.33 lbs/ft	
4 inch	4.50 od	0.120 in	4.26 id	0.237 in	4.026 id	0.337 in	3.826 id
Weight (lbs/ft.)	Steel	5.62 lbs/ft		10.80 lbs/ft		15.00 lbs/ft	
	Stainless						
	Aluminum	1.922 lbs/ft		3.730 lbs/ft		5.180 lbs/ft	

(Continued)

Table 6.1 *Cont.*

		Schedule (10, 40, 80) wall thickness (inches.)**					
		Sch.10		**Sch.40**		**Sch.80**	
Pipe size*	**O.D. (in.)**	**Wall (in)**	**I.D. (in)**	**Wall (in)**	**I.D. (in)**	**Wall (in)**	**I.D. (in)**
5 inch	5.563 od	0.134 in	5.295 id	0.258 in	5.047 id	0.375 in	4.813 id
Weight (lbs/ft.)	Steel	7.78 lbs/ft		14.63 lbs/ft		20.80 lbs/ft	
	Stainless						
	Aluminum	2.660 lbs/ft		5.050 lbs/ft		7.190 lbs/ft	
6 inch	6.625 od	0.134 in	6.357 id	0.280 in	6.065 id	0.432 in	5.761 id
Weight (lbs/ft.)	Steel	9.30 lbs/ft		18.99 lbs/ft		28.60 lbs/ft	
	Stainless						
	Aluminum	3.181 lbs/ft		6.560 lbs/ft		9.880 lbs/ft	
8 inch	8.625 od	0.148 in	8.329 id	0.322 in	7.981 id	0.500 in	7.625 id
Weight (lbs/ft.)	Steel	13.41 lbs/ft		28.58 lbs/ft		43.43 lbs/ft	
	Stainless						
	Aluminum			9.88 lbs/ft		15.010 lbs/ft	

*Nominal sizes apply – Pipe Size is the generic Industry Size Standard for reference only.

**Tolerances may vary slightly from each manufacturer

6.12 Stress

6.12.1 Radial Stress

The radial stress for a thick-walled pipe is given by the equation:

$$\sigma_r = [(p_i r_i^2 - p_o r_o^2)/(r_o^2 - r_i^2)] + [r_i^2 r_o^2 (p_o - p_i)/r^2 (r_o^2 - r_i^2)]$$

In this equation, r_i is the inner radius, r_o is the outer radius, p_i is the inner absolute pressure and p_o is the outer absolute pressure.

6.12.2 Tangential Stress

Tangential stress (also called hoop stress or circumferential stress) is the force exerted circumferentially (perpendicular both to the axis and to the radius of the object) in both directions on every particle in the cylinder wall. It can be described as:

$$\alpha\Theta = F/tl$$

F is the force exerted circumferentially on an area of the cylinder wall that has the following two lengths as sides: (i) t is the radial thickness of the cylinder and (ii) l is the axial length of the cylinder.

6.12.3 Longitudinal Stress

Longitudinal stress, also referred to as axial stress, can be calculated using the equation:

$$\sigma_a = F_T/[\pi(r_o^2 - r_i^2) + \sigma_{buck} + \sigma_b$$

Example calculation
Calculate the axial stress of the pipe with the following data:

- True axial force = 64.6 kips
- Effective force = 97.9 kips
- Pipe body diameter = 5 inches
- Pipe inside diameter = 4 inches
- Pipe data: premium class; assume no buckling and bending occur
- E = modulus of elasticity, psi; for steel = 30×10^6 psi

Solution:
Derating the pipe based on the premium class,

$$D_{cp} = c \times D_p + D_i(1\text{-}c)$$

$$\text{New OD} = 0.800 \times 5.0 + 4.00(1.0 - 0.800) = 4.8 \text{ inches}$$

The longitudinal stress inside the pipe is:

$$\Sigma a = FT/\pi(r_o^{\,2} - r_i^{\,2}) = 646000/\pi[(4.8/2)^2 - (4/2)^2] = 11{,}683 \text{ psi}$$

6.12.4 Stress Ratio

The stress ration is calculated from the equation:

$$X = (\sigma_y \times \%\ \text{yield})/\sigma_{vm}$$

where σ_y is the yield strength of the pipe in psi, and % yield is a percentage used to reduce the yield strength of the pipe as a factor of additional safety. When X is equal to or less than 1, there is concern of failure.

6.13 Tension

Tension (the opposite of compression) is the pulling force transmitted axially by means of a string, cable, chain, or similar one-dimensional continuous object, or by each end of a rod, truss member, or similar three-dimensional object. Effective tension is given by the relationship:

$$F_e = F_i + F_{bs}$$

$$Fe = F_i + P_oA_o + P_iA_i$$

6.14 Torque

Torque is rotational force and is defined as the cross product of the vector by which the force's application point is offset relative to the fixed suspension point (distance vector) and the force vector, which tends to produce rotational motion. The magnitude of torque depends on three quantities: (i) the force applied, (ii) the length of the lever arm connecting the axis to the point of force application, and (iii) the angle between the force vector and the lever arm. Thus:

$$\tau = r \times F$$

In the equation, r is the position vector (a vector from the origin of the coordinate system defined to the point where the force is applied) and F is the force vector. The magnitude $\boldsymbol{\tau}$ of the torque is given by the equation:

$$\tau = r.F \sin \theta$$

In this second equation, r is the distance from the axis of rotation to the particle, F is the magnitude of the force applied, and θ is the angle between the position and force vectors.

7

Drilling Tools

7.01 Backoff Calculations

The force at the backoff depth id equal to the axial force down to well depth minus the axial force down to backoff depth. Thus, the surface axial force when the workstring is at a measured depth (MD) is:

$$\Sigma MD \times (\text{weight gradient} + \text{drag force gradient})$$

The rotary table torque at surface is equal to the torque at backoff depth plus the backoff torque.

Example calculation

Estimate the surface action to backoff the string at a depth 9,000 ft with a backoff force and torque of 5 kips and 7,000 ft-lbf, respectively, applied at the backoff point. The string got stuck at 9,000 ft while tripping out. Other well and string details are given below:

- Well depth = 10,000 ft
- Wellbore details: casing with an inside diameter of 10" set at 6,000 ft and an open hole of 10-inch" diameter
- Drill string details:

Section type	Length (ft)	OD (in)	ID (in)	Weight (ppf)
Drill Pipe	7989	5	4	29
Drill Collar	540	8	2.5	154.3
Jar	30	8	3	147
Drill Collar	1440	8	2.5	154.3
Bit	1	10		95.37

- Drill pipe tool joint details: 7.25" OD × 3.5" ID
- Mud weight = 10.5 ppg
- Assume a friction factor of 0.3.
- The well is inclined 5° from the surface without any azimuth change.

Solution:

Buoyed weight calculation:

For the drill pipe, assume the tool joint length is 5% of the string length.

For the drill pipe, assume that the tool joint is 5% of the string length. Thus:

$$
\begin{aligned}
A_o &= \pi/490.95 \times D_p^{\ 2} + 0.05 \times D_{jt}^{\ 2}) \\
&= \pi/4(0.95 \times 5^2 + 0.05 \times 7.25^2)/144 \\
&= 0.14387 \text{ ft}^2
\end{aligned}
$$

$$
\begin{aligned}
A_i &= \pi/4(0.95 \times ID_p^{\ 2} + 0.05 \times ID_{jt}^{\ 2}) \\
&= \pi/4(0.95 \times 4^2 + 0.05 \times 3.5^2)/144 = 0.0862 \text{ ft}^2
\end{aligned}
$$

$$
\begin{aligned}
w_b &= w_s + p_i A_i - p_o A_o \\
&= 29 + 7.48052 \times 10.5(0.0862 - 0.14387) \\
&= 24.4737 \text{ ppf}
\end{aligned}
$$

Similarly, Similarly, the buoyed weight for the drill collar and jar are 129.56 ppf and 123.4381 ppf, respectively.

Side force calculation:
For the drill pipe, the side force of DP against the wellbore is equal to the component of the buoyed weight normal to the wellbore. Thus:

$$F_s = w_b \times \sin\alpha = 24.4737 \times \sin 5° = 2.133 \text{ ppf}$$

Similarly, the side force against the drill collar and jar are 11.29 ppf and 10.76 ppf, respectively.

Drag force calculation:
For the drill pipe, the drag force gradient is equal to the side force multiplied by the friction factor (coefficient of friction). Thus, the drag force gradient of DP against the wellbore is given by the relationship:

$$F_d = F_s \times \mu = 2.133 \times 0.3 = 0.64 \text{ ppf}$$

Similarly, the drag force gradient against the drill collar and jar are 3.39 ppf and 3.23 ppf, respectively.

Axial force at the backoff depth based on the trip out condition:
Force at the backoff depth is equal to the axial force down to the well depth minus the axial force down to the backoff depth.

Surface axial force down to the well depth is:

$$\Sigma MD \text{ (Weight Gradient + Drag Force Gradient).}$$

Thus:

$$7989(24.47 + 0.64) + 540(11.29 + 3.39) + 30(10.76 + 3.23) + 1440(11.29 + 3.39) = 230089.9 \text{ lbf}$$

Measured weight at well depth = 230,0090 + 50,000 = 280,090 lbf = 280 kips.

The axial force down to backoff depth:

$$= 7989(24.47 + 0.64) + 540(11.29 + 3.39) + 30(10.76 + 3.23) + 441(11.29 + 3.39) = 215425 \text{ lbf.}$$

Final surface axial force for backoff operation:
Measured weight at the surface for backing off is equal to the = axial force at backoff depth + backoff force + hoisting weight:

$$= 215425 + 5{,}000 + 50{,}000 = 270{,}424/1000 \text{ lbf} = 270 \text{ kips.}$$

The rotary table torque at the surface is equal to the torque at the backoff depth plus the backoff torque and the torque is equal to the component side force multiplied by the friction factor multiplied by the radius of the component.

The string torque at the backoff depth is:

$$7989[2.133 \times 0.3 \times 7.25/(2 \times 12)] + 540[11.29 \times 0.3 \times (8/(2 \times 12)] + 30[1076 \times 0.3(7.25/(2 \times 12] + 441[11.29 \times 0.3 \times 8/(2 \times 12)] = 2681$$

The surface torque for backoff = 2,681 + 7,000 = 9,681 ft-lbf.

7.02 Downhole Turbine

The torque developed is calculated from the relationship:

$$T = 2\pi Q\rho_m r^2 n_s N\eta$$

Q is the flow rate in gpm, ρ_m is the mud weight in ppg, r^2 is the square of the mean blade radius in square inches, ns is the number of turbine stages, N is the rotation speed of the turbine in rpm, and η is the efficiency.

Example calcualtion
Calculate the torque developed by a hydraulic turbine with a flow rate of 400 gpm and 10 ppg mud.

- Mean blade radius = 2.05 inches
- Number of stages = 100
- Rotation speed of turbine = 100 rpm
- Overall efficiency = 65%

Solution:
The torque developed is calculated as

$$T = 2\pi Q\rho_m r^2 n_s N\eta$$

Thus:

$$T = 2\pi \times 400 \times 10 \times 2.05^2 \times 100 \times 100 \times 0.65 \text{ ft lb}$$

7.03 Jar Calculations

A jar is an impact tool installed in the drillstring to free stuck pipe. A jar concentrates kinetic energy at the point where the pipe is stuck.

7.03.1 Force Calculations for Up Jars

The effective jar set (cock) force is given by the relationship:

$$Fes = -(Fs + F_{pof})$$

F_s is the set force, and F_{pof} is the pump open force.
The effective jar trip force is given by

$$F_{et} = (F_{ts} - F_{pof})$$

F_{ts} is the trip force. An Induced force is required to set the jar at the center of the jar in compression, and this requires over pulling the measured weight (hoisting/trip out). The set measured weight is given by:

$$F_{emw} = (F_{ti} - F_s - F_{pof})$$

F_{tis} is the trip in axial force, and F_s is the up jar set force.
An induced force is also required to trip the jar at the center of the jar in tension, and this requires slacking off the measured weight (lowering/trip in). The trip measured weight is given by the relationship:

$$F_{tmw} = (F_{to} + F_t - F_{pof})$$

F_{to} is the trip out axial force, and F_t is the up-jar trip force.

7.03.2 Force Calculations for Down Jars

The effective jar set (cock) force is calculated from the relationship:

$$F_{es} = (F_s - F_{pof})$$

The effective jar trip force is calculated from:

$$F_{es} = (F_{ts} - F_{pof})$$

An induced force is required to set the jar at the center of the jar in tension, and this requires over pulling the measured weight (hoisting/trip out). The set measured weight is given by the equation:

$$F_{emw} = (F_{to} + F_{s} - F_{pof})$$

An induced force is required to set the jar at the center of the jar in tension, and this requires over pulling the measured weight (hoisting/trip out). Set measured weight is given by the relationship:

$$F_{tmw} = (F_{i} - F_{t} - F_{pof})$$

Example calculation
Calculate the jar forces using the following data:

- Well depth = 8,000 ft
- Drill pipe = 7000 ft 5 inches, 4 inches OD, ID = 19.5 lb/ft (ppf), S = 105
- Drill collars = 500 ft, 8 inches OD, 3inches ID
- Jar = 30 ft, 8 inches OD, 3 inches ID
- Drill collars = 500 ft, 8 inches OD, 3 inches ID
- Bit = 12.25 inches
- Casing = 5000 ft, 13.375 inches
- Open hole = 12.25 inches
- Hoisting equipment weight = 30 kips
- Trip in axial force at the center of the jar = 150 kips
- Trip out axial force at the center of the jar = 200 kips
- Pump open force = 3000 lbs
- Type of jar: mechanical
- Up jar set (cock) force = 40 kips
- Down jar set (cock) force = 30 kips
- Measured weight when stuck = 350 kips

Solution:
Up jar conditions:

Set (cock) measured weight at the surface = 150 – 40 – 3 + 30 = 137 kips.

Change in measured weight = 137 – 350 = -213 kips

The trip measured weight at the surface = 200 + 40 – 3 + 30 = 267 kips.

The change in set measured weight = 267 – 137 = –130 kips.

Reset (re-cock) measured weight at the surface = 137 kips.

The change in reset measured weight from the trip measured weight = 137 – 267 = -130 kips.

Down jar conditions:

Set (cock) measured weight at the surface = 200 +30 – 3 + 30 = 257 kips.

Measured weight = 350 – 200 – 50 = 100 kips.

Since the measured weight is less than the set measured weight, the jar would have already set. For the down jar, for setting, the string must be tripped out to 257 kips, but the present measured weight in stuck conditions is 350 kips.

The change in measured weight = 350 – 350 = 0 kips.

The trip measured weight at the surface = 150 – 40 – 3 + 30 = 137 kips.

The change in set (cock) measured weight = 137 – 350 = –213 kips. Reset (re-cock) measured weight at the surface = 200 + 30 – 3 + 50 = 277 kips.

The change in reset measured weight from the trip measured weight = 277 – 137 = 140 kips.

7.04 Overpull/Slack-off Calculations

When the pipe is s tuck, if the surface measured weight at the stuck condition is known, then

$$F_{sp} = F_{td} - F_{sd}$$

F_{sp} is the force at stuck point depth, F_{td} is the axial force down to well depth, and F_{sd} is the axial force down to stuck depth.

When the pipe is stuck, if the surface measured weight at the stuck condition is known, then

$$F_{sp} = F_{td} - F_{sd}, \tag{7.4}$$

where F_{sp} = force at stuck point depth, F_{td} = axial force down to well depth, and F_{sd} = axial force down to stuck depth.

Problem 7.3

Estimate the surface action to load the stuck point depth of 9,100 ft.

Use the data from Problem 7.2.

Solution:

Axial force at the stuck point depth based on the trip out condition:

Force at stuck point depth = axial force down to well depth – axial force down to backoff depth

The axial force down to well depth is

$$F_{td} = 7989(28.48 + 0.64) + 540(11.29 + 3.39) + 30(10.76 + 3.23)$$
$$+1440(11.29 + 3.39)$$
$$= 230{,}089 \text{ lbf.}$$

The axial force down to the stuck points

$$7989(24.47 + 0.64) + 540(11.29 + 3.39) + 30(10.76 + 3.23) + 540(11.29 + 3.39) = 216{,}878 \text{ lbf.}$$

The force at the stuck point depth

$$230{,}089 - 216{,}878 = 13{,}212 \text{ lbf} = 13.2 \text{ kips.}$$

Measured weight

$$= 338{,}790 + 50{,}000 = 388{,}790 \text{ lbf} = 389 \text{ kips.}$$

The axial force at the stuck point depth based on the trip in condition is calculated as

$$F_{td} = 7989(24.47 - 0.64) + 540(11.29 - 3.39) + 30\,(10.76 - 3.23) + 1540(11.29 - 3.39) = 207{,}035 \text{ lbf.}$$

7.05 Percussion Hammer

For the percussion hammer, the work done (WD) is calculated by the equation:

$$WD = (\Delta p_m \times A_p \times L \times nb)/396{,}000 \text{ hp}$$

In this equation, p_m is the pressure drop across the piston chamber in psi, A_p is the cross-sectional area of the piston in square inches, L is the stroke length in inches, and n_b is the number of blows of the piston per minute.

Example calculation
Calculate the work done in HP by a percussion hammer with the following details:

- Pressure drop in the chamber = 400 psi
- Diameter of the pressure changer = 4 inches
- Stroke length = 1 foot
- Number of blows per minute = 200

$$WD = (400 \times \pi/4 \times 12 \times 200)/396{,}000 = 30.46 \text{ hp}$$

7.06 Positive Displacement Motor (PDM)

The mechanical developed by the positive displacement motor (MHP) is given by the relationship:

$$MHP = [(T/550)/(2\pi/60)] \times N$$

N is the rpm.

7.07 Rotor Nozzle Sizing

The following simple steps help to size the nozzle effectively:

1. Establish the differential pressure range based on the expected weight on the bit range.

2. Calculate the range of operating flow rates, *Qop*, required for the run.
3. Estimate the minimum flow rate required for hole- cleaning, *Qmin*.
4. If the operating flow rate is less than the minimum flow rate for hole-cleaning, calculate the additional flow rate, *Qrn*, that will be bypassed through the rotor nozzle.
5. Size the nozzle using the equation:

$$A_{rn} = (8.311 \times 10^{-5} \times Q_{rn}^{\ 2} \times \rho_m)/(C_d^{\ 2} \times \Delta p_m)$$

   A_{rn} is the area of the rotor nozzle in square inches, Q_{rn} is the bypass flow rate through the rotor nozzle in gpm, ρ_m is mud density of the circulating fluid in ppg, C_d is the discharge coefficient, Δpm is the pressure drop across the motor in psi. The proper nozzle size can be calculated by rearranging the above equation. The rotor nozzle is often expressed in 32^{nds} of an inch. If the rotor nozzle is specified as "14," then the rotor nozzle has a diameter of 14⁄32 of an inch.
6. Check that the diameter of the nozzle is sufficiently smaller than the shaft diameter.

Example calculation

A well is planned to be drilled with an 8.5-in class 1-1-1 bit, while the torque and rpm expected are 3000 ft-lbf and 300 rpm, respectively. The mud weight required is 10 ppg. Determine the size of the rotor nozzle for the following conditions:

Minimum flow rate required for hole-cleaning = 475 gpm

- Configuration: 2/3
- Diameter of the motor = 6.75 inches
- Pitch of the housing = 23 inches

Solution:

Diameter of the housing is assumed to be 6 inches. Thus, the pressure drop to be expected across the motor power section is:

$$\Delta p_m = 3000/[0.01 \times 0.66(1.66/2\text{-}0.66632 \times 23 \times 26 \times 0.7)]$$

$$= 836 \text{ psi}$$

The operating flow rate requited is:

$$Q_{op} = (300 \times 0.79 \times 0.666 \times 1.666 \times 23 \times 36)/(230.98 \times 1.333^2) = 530 \text{ gpm}$$

Since this flow rate is higher than the minimum required flow rate of 475 gpm, there is no necessity to fit a rotor nozzle.

7.08 Stretch Calculations

Based on the pipe stretch under the applied tension, the approximate depth of the stuck point can be derived using the equation:

$$L = (E \times e \times W)/(144 \times \Delta T \times \rho_s)$$

E is Young's modulus in psi, e is the measured elongation corresponding to the differential tension (pull) in inches, W is air weight of the pipe in lb, ΔT is the differential pull or hook load in inches, and ρ_s is the density of steel in lb/in^3. For steel, the above equation can be written as:

$$L = (735294 \times e \times W)/\Delta T$$

Example calculation
A driller is planning to estimate the approximate stuck point in a vertical well by conducting the stretch test. The following is the observed data:

- Initial pull = 210 kips
- Final pull = 240 kips
- Hook loads are greater than the weight of the string.
- Stretch of the pipe observed between these hook loads is found to be 19 inches.
- Drill pipe: 5 inches, 19.5 ppf, IEU

Calculate the estimated depth of the stuck point.

Using the above equation, the depth of the stuck point is derived from the equation above. Thus:

$$L = (735294 \times 19 \times 19.5)/(30 \times 1{,}000) = 9{,}080 \text{ feet}$$

8

Pore Pressure and Fracture Gradient

8.01 Formation Pressure

Formation pressure arises from the pressure of fluids within the pores of a reservoir, usually hydrostatic pressure, or the pressure exerted by a column of water from the formation's depth to sea level. When a formation is impermeable, such as is the case in a shale formation, the pore fluids cannot always escape and must then support the total overlying sediments which leads to high formation pressures. Because reservoir pressure changes as fluids are produced from a reservoir, the pressure should be described as measured at a specific time, such as initial reservoir pressure.

8.01.1 Hubert and Willis Correlation

The matrix stress is calculated from the relationship:

$$\sigma z = \sigma\ ob - pp\ psi$$

In this equation, where σob is the overburden pressure in psi, and pp is the pore pressure, also psi.

The fracture pressure is obtained from the relationship:

$$p_{ff} = \sigma_h + p_p \text{ psi}$$

$$p_{ff} = \sigma_{min} + p_p \text{ psi}$$

σ_h is the horizontal stress in psi and σ_{min} is the minimum horizontal stress, also in psi. The Horizontal stress is assumed to be one-half and one-third of the overall stress. Thus:

$$\sigma_{hmin} = 0.33\sigma_z = 0.33(\sigma_{ob} - p_p) \text{ psi}$$

$$\sigma_{hmax} = 0.5\sigma_z = 0.5(\sigma_{ob} - p_p) \text{ psi}$$

Thus, the fracture pressure is calculated from the relationship:

$$p_{ff} = 0.33\sigma_z = 0.33(\sigma_{ob} - p_p) \text{ psi}$$

$$p_{ff} = 0.5\sigma_z = 0.5(\sigma_{ob} - p_p) \text{ psi}$$

In terms of pressure gradient:

$$p_{ffmin}/D = 0.33[(\sigma_{ob} - p_p)/D] \text{ psi}$$

$$p_{ff}/D = 0.5[(\sigma_{ob} - p_p)/D] \text{ psi}$$

D is the depth in feet.

8.01.2 Matthews and Kelly Correlation

The minimum pressure required to create a fracture is at least the formation fluid pressure, and any additional pressure may be related to overcome the formation matrix:

$$\sigma_z = \sigma_{ob} - p_p \text{ psi}$$

$$\sigma_{min} = F_s\sigma_z \text{ psi}$$

$$p_{ff} = F_\sigma\sigma_z + p_p \text{ psi}$$

$$p_{ff}/D = (F_\sigma \sigma z)/D + p_p/D \text{ psi/ft}$$

F_σ is the matrix coefficient

8.01.3 Eaton's Correlation

The horizontal and vertical stress ratio and the matrix stress coefficient are dependent on the Poisson's ratio of the formation.

$$\sigma_x + \sigma_y + \sigma_h = [\mu/(1 - \mu)]\sigma_z \text{ psi}$$

$$p_{ff} = [\mu/(1 - \mu)](\sigma_{ob} - p_p) + p_p \text{ psi}$$

$$p_{ff}/D = [\mu/(1 - \mu)][(\sigma_{ob} - p_p)/D] + p_p/D \text{ psi/ft}$$

In these equations, μ is the Poisson ratio.

8.01.4 Christman's Correlation

The effect of the water depth in calculating the overburden gradient is accounted for by the relationship:

$$p_{ff}/D = (1/D)(\rho_w D_w + \rho_b D_f) \text{ psi/ft}$$

If it is assumed that the density of sea water is 1.02 gm/cc, the equation becomes:

$$p_{ff}/D = (1/D)(0.44D_w + \rho_b D_f) \text{ psi/ft}$$

8.02 Leak-off Pressure

Leak-off occurs when the fracture area is constant during shut-in and the leak-off occurs through a homogeneous rock matrix. The formation breakdown gradient is equal to the leak-off pressure minus the true vertical depth (TVD) of the shoe). Thus, the equivalent mud weight to fracture the formation is derived from the equation:

$$\rho_{ff} = p_s/(0.052 \times D_v) + \rho_{test} \text{ ppg}$$

ps is the surface leak-off pressure during the test in psi, Dv is the true vertical depth of the formation tested in feet, and ptest is the mud weight during the test in ppg. The maximum fracture pressure equivalent mud weight required to drill the next formation interval is derived from the equation: $(\rho_{ff} - \rho_{test})0.052 \times D_v$ in ppg.

Example calculation

Calculate the mud weight required to break down the formation with the following data taken during a leak-off test.

- Surface pressure during the leak-off = 500 psi
- Mud weight during the test = 10 ppg
- True vertical depth of the formation tested = 5002 feet

Solution:

The equivalent leak-off mud weight

$$\rho_{ff} = \rho_s/(0.052 \times D_v) + \rho_{test} \text{ ppg}$$

$$= 500/(0.052 \times 5002) + 10 = 11.922 \text{ ppg}$$

The formation pressure at the leak-off depth derives using the equation is:

$$Pff = 0.052 \times 11.922 \times 5002 = 3{,}101 \text{ psi}$$

9

Well Control

9.01 Accumulators

In the drilling technology area, a hydraulic accumulator is a pressure storage reservoir in which a non-compressible hydraulic fluid is held under pressure that is applied by an external source. An accumulator enables a hydraulic system to cope with extremes of demand using a less powerful pump, to respond more quickly to a temporary demand, and to smooth out pulsations. Compressed gas accumulators, also called hydro-pneumatic accumulators, are by far the most common type of accumulators.

An accumulator follows a specific charging sequence. Thus:

The volume delivered by the accumulator bottles is given by the formula:

$$V_d = (P_{pc}/P_f - P_{pc}/P_s) \times V_b$$

P_{pc} is the nitrogen pre-charge pressure in p_{si}, P_f is the final operating pressure of the bottle in p_{si}, P_s is the accumulator system pressure in p_{si}, and V_b is the actual bottle capacity in gallons.

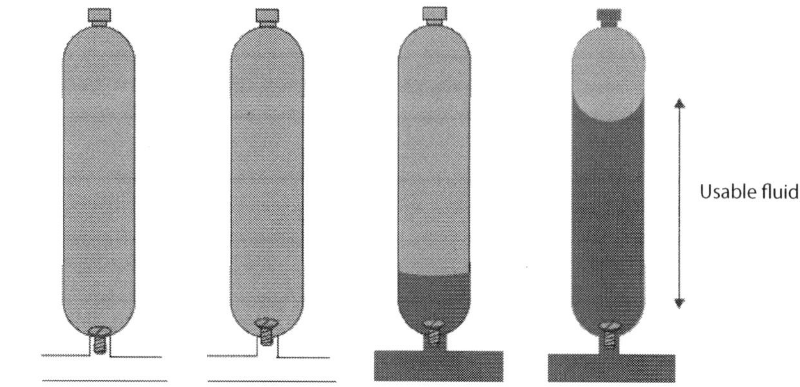

Figure 9.1 Charging Sequence for an Accumulator.

For sub-sea operations, the volume delivered is given by the equation:

$$V = Vd[(460 + T_2)(460 + T_1)]$$

where T_1 is the ambient temperature in degrees Fahrenheit and T2 is temperature at water depth, also in degrees Fahrenheit.

Example calculation
Compute the total amount of fluid delivered from eight 20-gallon bottles with a system pressure of 3,000 psi. The pre-charge pressure is 1100 psi, and the final pressure is 1250 psi.

Solution:
The volume delivered is calculated from the equation:

$$V_d = (P_{pc}/P_f - P_{pc}/P_s) \times V_b = 8 \times [(1100/1250 - 1100/3000)] \times 20$$
$$= 81.6 \text{ gallons}$$

9.02 Driller's Method

Assuming a gas kick, during the first circulation the kick fluid is circulated out of annular space using the original light mud with- out allowing any further kick fluid intrusion by keeping the circulating drill pipe pressure at:

$$Pcdpi = Ppr + Psidp.$$

Afterwards, close the choke and read Psidp and Psic and the two should be equal. If not, continue circulating until all gas is removed from the well.

During the second circulation, a volume of the heavy kill mud that is equal to the total drillstring volume capacity is circulated while keeping the choke pressure at Psidp. The circulating drill pipe pressure should drop from Pcdpi at the beginning of the circulation to Pcdpf when the new mud reaches the bit. Circulation of heavy mud is continued until all original mud in the annulus is displaced. Afterwards, close the choke and read Psidp and Psic and the two should be equal to 0 psi. If so, open the well and resume the drilling operations. If not, continue the well control operation.

9.03 Formulas Used in Kick and Kill Procedures

9.03.1 **Hydrostatic Pressure** (psi): MW × TVD × 0.0519
MW is the Mud Density (lb/gal)
TVD is the True Vertical Depth (ft)

9.03.2 **Circulating Pressure** (psi): (MW × TVD × 0.0519) + Pla
Pl_a is the Annular Pressure Loss (psi)

9.03.3 **Initial Circulating Pressure** (psi): SPR + SIDP
SPR is the System pressure loss at kill rate (psi)
SIDP is the Shut-in Drill Pipe Pressure (psi)

9.03.4 **Final Circulating Pressure** (psi): (KMW/MW) × SPR
KMW = Kill Mud Density (lb/gal)

9.03.5 **Kill Mud Weight** (lb/gal): MW + (SIDP / (TVD × 0.0519))

9.03.6 **Formation Pressure** (psi): SIDP + (MW × TVD × 0.0519)

9.03.7 **Density of influx** (ppg): MW – [(SICP – SIDP)/(L × 0.0519)]
SICP is the Shut-in casing pressure (psi)
L is the Length of influx (ft)

9.03.8 **Length of kick around drill collars** (ft):
Pit Gain* (bbls)/Annular Volume around collars (bbls/ft)

9.03.9 **Length of kick, drill collars and drill pipe** (ft):
Collar Length + [(Pit Gain* – Collar Annular Volume)/(D_1^2 – D_2^2 × 0.000971)]
D_1 is the hole diameter (inches)
D_2 is the drill pipe diameter (inches)

9.03.10 **Gas bubble migration rate** (psi/hr): ΔPa / (0.0519 × MW)
ΔPa is the pressure change over time interval/time interval (hr)

9.03.11 **Barite required** (sk/100 bbls mud): 1490 × (KMW – MW) / (35.8 – KMW)

9.03.12 Volume increase caused by weighting up: 100 × (KMW – MW) / (35.8 – KMW)

*Fluids entering the wellbore displace an equal volume of mud at the flowline, resulting in pit gain.

9.04 Hydrostatic Pressure Due to the Gas Column

The surface pressure due to the gas column is given as

$$Pg = Poe^{(0.01877 \times \rho g \times D)/ZT}$$

Po is the formation pressure of gas in psi, ρg is the density of the gas in ppg, D is the height of the gas column in feet, T is the temperature in degrees Kelvin, Z is the compressibility factor at temperature T, and P – Po the hydrostatic pressure in psi.

Example calculation

A 10,000 ft well full of gas was closed at the surface. The formation pressure and temperature is 6,200 psi and 180°F, respectively. Assume a gas gravity of 0.7 and Z = 0.9 and a temperature equal to 180 + 460 = 640°K.

The surface pressure due to the gas column is given by the relationship:

$$Pg = Poe^{(0.01877 \times \rho g \times D)/ZT} = 6200e^{(0.0187 \times 0.7 \times 10000)/(0.9 \times 630)} = 7788.57 \text{ psi}$$

The hydrostatic pressure exerted due to the gas column is:

$$Pg - Po = 7788.57 - 6200 = 1589 \text{ psi}$$

9.05 Kill Methods

The two widely used constant bottomhole circulating methods are the Driller's Method and the Wait and Weight (W&W) Method. The basic principle of both methods is to keep bottomhole pressure (BHP) constant at or, preferably, slightly above the formation pressure. Relevant formulas are:

1. **Kill Weight Mud**

$$\mathbf{KWM = OMW + [SIDPP \div (0.052 \times TVD)]}$$

KWM is the kill weight mud in ppg, OMW is the original mud weight in ppg, SIDPP is shut in drill pipe pressure in psi, TVD is the true vertical depth of the well in feet.

2. **Slow Circulating Rate (SCR)**

$$\mathbf{SCR = ICP - SIDPP}$$

SCR is the slow circulating rate in psi, ICP is the initial circulating pressure in psi, and SIDPP is the shut-in drill pipe pressure in psi.

3. **Final Circulating Pressure (FCP)**

$$\mathbf{FCP = SCR \times KWM \div OMW}$$

FCP is the final circulating pressure in psi, SCR is the slow circulating rate in psi, and KWM is the kill weight mud in ppg.

9.06 Kill Mud Weight

The formation pressure is calculated from the relationship:

$$P_f = 0.062 \times \rho_m \times D_v + P_{sidp}\ \text{psi}$$

Thus, the kill mud weight, K_m, is:

$$K_m = \rho_m + [P_{sidp}/(0.052 \times D_v)] + \rho_{ok}\ \text{psi}$$

In this equation, ρm is the original mud weight in ppg, Psidp is the shut-in drill pipe pressure in psi, Dv is the vertical depth in feet, and ρok is the overkill safety margin in ppg.

The initial circulating pressure (ICP) is:

$$\text{ICP} = \text{Psidp} + \text{Pp} + \text{Po}$$

Psidp = the shut-in drill pipe pressure in psi, Pp is the slow circulating pump pressure in psi, and Po is the overkill pressure, psi. Thus, the final circulating pressure (FCP) is:

$$\text{FCP} = P_p \times (\rho_{km}/\rho_{om})$$

Example calculation

Estimate the kill fluid density for a shut-in-drill pipe pressure of 580 psi. The kick depth is 11,937 feet and the original mud density is 14.3 ppg.

Solution:

$$\text{Kill fluid density} = 14.3 + 580/(0.052 \times 11937) = 15.3 \text{ ppg}$$

$$\text{Kill fluid gradient} = 14.3 \times 0.052 + 480/11937 = 0.7922 \text{ psi/foot.}$$

9.07 Leak-off Pressure

Typically leak-off occurs when the fracture area is constant during shut-in and the leak-off occurs through a homogeneous rock matrix. The leak-off pressure, also often referred to as the formation breakdown pressure, is calculated from the relationship:

$$P_{ff} = 0.052 \times D_v \times \rho_m + P_{lop} \text{ psi}$$

D_v is the true vertical depth in feet, ρ_m is the mud density, and P_{lop} is the surface leak-off pressure.

The equivalent mud gradient is calculated using the relationship: P_{ff}/D_v psi/foot and the formation breakdown gradient, FBG, is calculated as:

$$FBG = (P_{ff} - D_v \times \rho_{sw})/(D_v - D_a - D_s)$$

D_v is the true vertical depth in feet, ρ_{sw} is the sea water density in ppg, D_a is the air gap in feet, and D_s is the sea depth in feet.

Example calculation

A well planner calculates the equivalent fracture gradient using the following relationship:

$$\rho_{ff} = [F_s(\sigma_{ob} - 0.052) - \rho_f] + \rho_f$$

The overburden gradient, σ_{ob}, is $0.89934e^{0.0000Dv}$ and the matrix stress ratio, F_s, is calculated as:

$$Fs = 0.327279 + 6.94721 \times 10^{-5}Dv - 1.98884 \times 10^{-9}D_v^2$$

Calculate the formation gradient at a depth 9.875 ft.

Solution:

The matrix stress ratio at 9,875 ft is calculated from the equation:

$$Fs = 0.327279 + 6.94721 \times 10^{-5}9875 - 1.98884 \times 10^{-9}9875^2 = 0.823$$

The overburden gradient is:

$$\sigma_{ob} = 0.89934e^{0.0000x9875} = 0.9938 \text{ psi/foot}$$

The formation fracture gradient is calculated as

$$\rho_{ff} = 0.823(0.9938/0.052 - 9) + 9 = 17.32 \text{ ppg}$$

9.08 Length and Density of the Kick

A kick is an undesirable influx of formation fluid into the borehole. If left unattended, a kick can develop into a blowout (an uncontrolled influx of formation fluid into the borehole). Failure to control a kick can be the loss of the well and serious consequences for the rig and the health and welfare of the crew.

9.08.1 Length of the Kick

Length of the kick, ft = pit gain, bbl/(annulus capacity behind drill collar, bbl/ft)

For pit gain, bbl < annulus volume against drill collar, bbl:

Length of the kick, ft = length of the drill collar, ft + (pit gain, bbl – annulus volume against drill collar, bbl)/annulus capacity behind drill pipe, bbl/ft

For pit gain, bbl > annulus volume against drill collar, bbl

9.08.2 Density of the Kick

Density of the kick, ppg = initial mud weight, ppg – (initial stabilized drill pipe pressure in psi initial stabilized casing pressure, psi)/(0.052 × Length of the kick, ft)

9.08.3 Type of Kick

A gas kick is given by: $\rho_k < 0.25$ psi/ft
An oil and gas mixture is given by: $0.25 < \rho_k < 0.3$ psi/ft.
An oil or condensate is given by: $0.3 < \rho_k < 0.4$ psi/ft
A water kick is given by: $0.4 < \rho_k$ psi/ft

9.08.4 Kick Classification

The kick while drilling is classified using the following criteria:

Pore pressure > dynamic bottomhole pressure > static bottomhole pressure

The kick after pump shutdown is classified by the following criteria:

Dynamic bottom hole pressure > pore pressure > static bottomhole pressure

The swab kick is classified by the following criteria:

Dynamic bottomhole pressure > static bottomhole pressure > sore pressure

Example calculation
A kick was taken when drilling a high-pressure zone of a depth of 9,875 ft with a mud density of 9 ppg. After the well was shut-in, the pressures recorded were SIDP = 300 psi and SICP = 370 psi. The total pit gain observed was 5 bbl. The annular capacity against 900 ft of drill collar is 0.0292 bbl/ft. The overkill safety margin is 0.5 ppg. Calculate the formation pressure, kick density, the type of fluid, and the required kill mud weight.

Solution:
The formation pressure, P_f, is calculated from the equation:

$$P_f = 0.052 \times \rho_m \times D_v + P_{sidp} = 0.052 \times 9 \times 9875 + 300 = 4921 \text{ psi}$$

Total annular volume against the drill collar is equal to $0.0292 \times 900 = 26.28$ bbl. Therefore, the volume of the kick is

$$V_k < V_{an/dc}$$

Therefore, the length of the kick is:

$$L_k = V_k/C_{an/dc} = 5/0.0292 = 171.23 \text{ ft}$$

The density of the kick fluid is:

$$\rho_k = \rho_m = (P_{sidp} - P_{sicp})/(0.052 \times L_k) = 9 + (300 - 370)/0.052 \times 171.23) = 1.13 \text{ ppg}$$

Therefore, the kick fluid is gas.
The kill mud weight is:

$$\rho_{km} = \rho_m + P_{sidp}/(0.052 \times D_v) + \rho_{ok} = 9 + 300/(0.052 \times 9875 + 0.5 = 10.08 \text{ ppg})$$

9.08.5 Kick Tolerance

In terms of kick tolerance:

1. If $D_v - L_k < D_{cs}$, then:

$$K_T = (\rho_{ff} - \rho_k)(D_{cs}/D_v) + \rho_k - \rho \text{ ppg}$$

2. If $D_v - L_k > D_{cs}$, then:

$$K_T = (\rho_{ff} - \rho)(D_{cs}/D_v) + (\rho - \rho_k)(L_k/D_v) \text{ ppg}$$

where ρ_{ff} is the formation fracture gradient at the shoe in ppg, Dcs is the vertical depth of the shoe in feet, ρ_k is the density of the kick in ppg, ρ is the density of the wellbore mud at the time of the kick in ppg, and L_k is the length of the kick.

Example calculation
Determine the kick tolerance for the following data:

- Well depth = 9,878 ft
- Casing shoe depth = 6,500 ft
- Mud density = 10.1 ppg
- Equivalent fracture mud weight at the shoe = 14.8 ppg
- Volume of the kick = 10 bbl
- Annulus capacity of 1,080 ft of drill collar = 0.0292 bbl/ft
- Kick density = 2 ppg

Solution:

The total annular volume against the drill collar is 0.0292 × 1080 = 31.536 bbl.

Therefore, the volume of the kick is: $V_k < V_{an/dc}$

The length of the kick, L_k, is calculated is:

$$L_k = V_k/C_{an/dc} = 10/0.0292 = 342.46 \text{ feet}$$

Calculating $D_v - L_k = 9878 - 342.46 = 9535.53$, which is $> D_{cs} = 6500$ft. Therefore, using the second condition:

$$K_T = (\rho_{ff} - \rho)(D_{cs}/D_v) + (\rho - \rho_k)(L_k/D_v)$$

$$= 14.0 - 10.1(6500/9878) + (10.1 - 2)(342.46/9878) - 3.37$$

9.09 Maximum Allowable Annular Surface Pressure

The maximum allowable annular surface pressure (MAASP) (static) is given by the equation:

$$P^s_{maasp} = 0.052(\rho_f - \rho_m)/D_{sv} \text{ psi}$$

D_{sv} is the shoe depth (TVD) in feet, ρ_f is the equivalent fracture mud weight in ppg, and ρ_m is the wellbore mud weight in ppg.

The maximum allowable annular surface pressure (dynamic) is given by the equation:

$$P^d_{maasp} = 0.052(\rho_f - \rho_m)/D_{sv} - \Delta_{pa} \text{ psi}$$

Δ_{pa} is the annular frictional pressure loss above the shoe in psi.

Example calculation

A well is being drilled with a mud weight of 14 ppg at a depth of 15,000 ft (TVD) has a shoe at 13,100 ft (TVD). The fracture gradient at the shoe is 0.85 psi/ft. Calculate the maximum allowable annular surface pressure.

Solution

This is a static condition. Therefore, the equivalent fracture mud weight, ρ_f, is:

$$\rho_f = 0.85/0.52 = 16.35 \text{ ppg}$$

The maximum allowable annular surface pressure (static) is:

$$P^{s}_{maasp} = 0.052(16.35 - 14)13100 = 1598psi$$

9.10 Riser Margin

Riser margin is equal to (drilling fluid gradient to control the formation pressure with riser in psi/foot multiplied by the depth of the hole (TVD) in feet minus the seawater gradient in psi/foot. The seawater gradient in psi/foot is the water depth in feet divided by the depth of the hole (TVD) in feet minus the water depth in feet. Thus, and equation by which the riser margin can be calculated is:

$$\rho_{rm} = (\rho_{emw} D_{v} - \rho_{w} D_{w})(D_{v} - D_{w} - D_{a})$$

In this equation, ρ_{emw} is the drilling fluid gradient to control the formation pressure with the riser in psi/ft, D_v is the depth of the hole (TVD) in feet, ρ_v is the seawater gradient, psi/ft, D_w is the water depth in feet, and D_a is the air gap, feet.

Example calculation
Calculate the riser margin for the given data: (i) water depth = 3575 ft, (ii) air gap = 75 ft, (iii) depth of the hole = 15,554 ft (TVD), and (iv) the formation gradient at the depth of hole is 0.78 psi/ft.

Solution:
The riser margin is:

$$(0.78 \times 15554 - 0.434 \times 3575)/(15554 - 3575 - 75) = 0.89psi/ft$$

The safety riser margin = 0.89 – 0.78 = 0.11 psi/ft

10

Drilling Problems

10.01 Differential Sticking Force

Differential sticking – a problem that occurs worldwide in terms of time and financial cost – is a condition whereby the drill string cannot be moved (rotated or reciprocated) along the axis of the wellbore and typically occurs when high-contact forces caused by low reservoir pressure, high wellbore pressure, or both, are exerted over a sufficiently large area of the drillstring. Differential sticking is a product of the differential pressure between the wellbore and the reservoir and the area that the differential pressure is acting upon. This means that a relatively low differential pressure (Δp) applied over a large working area can be just as effective in sticking the pipe as can a high differential pressure applied over a small area.

The formula for the approximate calculation of the differential sticking force due to fluid filtration is:

$$F_{pull} = \mu \Delta P A_c$$

ΔP is the differential pressure in psi ($\Delta P = P_m - P_f$), A_c is the contact surface area in square inches, P_m is the pressure due to the drilling fluid in psi, and P_f is the formation pressure or pore pressure in psi. The mud pressure is given by the expression:

$$P_m = 0.052 \times \rho_m \times D_v \text{ psi},$$

where ρ_m is the mud weight in ppg, and D_v is the vertical depth of the calculation or stuck depth, ft.

10.01.1 Method 1

The contact area is given by:

$$A_c = 2 \times 12 \times L_p\{[(Dh/2 - t_{mc})^2] - [(D_h/2 - t_{mc} \times (Dh - t_{mc}))/(D_h - D_p))]^2\}^{1/2} \text{ in}^2$$

D_h is the hole diameter in inches, D_p is the outer pipe diameter in inches, t_{mc} is the mud cake thickness in inches, L_p is the embedded pipe length in feet, D_{op} must be equal to or greater than $2t_{mc}$ and equal to or less than $(D_h - t_{mc})$.

10.01.2 Method 2

The contact ares is given by the equation:

$$A_c = Lp(D_pD_h)/(D_h - D_p)^{1/2}[(1 - X/X)(Qf_{tmc}/A_{f)}(\Delta p/\Delta p_f)(\mu_f\mu_{df})(t/t_f)]^{1/4}$$

Q_g is the measured filtrate volume, A_f is the filtration area, Δp_f is the filtration test differential pressure, μ_f is the filtrate viscosity under the test conditions, μd_f is the filtrate viscosity at downhole conditions, X is the contact ratio, t is the total test time, and t_f is the total stuck time.

Example calculation

Determine the pullout force to free the drill string given the following well data: (i) the OD of the drill collar = 6.0 inches, (ii) the hole size = 9.0 inches, (iii) the mud cake thickness = 1/16th (0.0625) of an inch, (iv) the coefficient of friction = 0.15, (v) the length of the embedded drill collar = 20 feet, and (vi) the differential pressure = 500 psi.

Solution

$$\mu = 0.15.\ \Delta P = 500 \text{ psi}.\ L_p = 20 \text{ ft}.$$

Thus:

$$A_c = 2 \times 20 \times 12[(9/2 - 0.0625)^2 - (9/2 - 0.0625)(9 - 6)^2]^{1.2}$$
$$= 500.4 \text{ square inches}$$

Therefore, pull out force = $\mu \Delta P A_c = 0.15 \times 2.085 \times 500 = 37{,}530$ lbf

10.01.3 Method 3

The arc length at the contact point is derived using the equation:

$$A_c = (\alpha/2)D_p \times L_p$$

$$\sin \alpha = [\varepsilon(X + \varepsilon)(X + D_p + \varepsilon)D_p - \varepsilon)]^{1/2}/[(X/2 + \varepsilon)D_p]$$

X is equal to $D_h - D_p - 2t_{mc}$, α is the angle of contact between the bottom hole assembly (BHA) section and the mud cake in radians, and ε is the deformation of the mud cake at the mid-point of contact in inches.

10.02 Hole Cleaning–Slip Velocity Calculations

The slip velocity arises because of the difference between the average velocities of two different fluids flowing together in a pipe. In vertical ascending flow, the lighter fluid flows faster than the heavier fluid. The slip velocity depends mainly on the difference in density between the two fluids, and their respective hold up.

10.02.1 Chien Correlation

The slip velocity if given for two conditions as illustrated in the following equations:

$$v_s = 0.458\beta\{[36800d_s/\beta_2)(\rho_s - \rho_m/\rho_m)]^{1/2} + 1 - 1\} \text{ (for } \beta < 10)$$

$$v_s = 86.4d_s(\rho_s - \rho_m/\rho_m)]^{1/2} \text{ (for } \beta > 10)$$

In these equations, $\beta = \mu_a/\mu_{mds}$ and $\mu_a = \mu_p + (300\tau_y d_s/V_a)$ where ρm is the mud density in ppg, ρ_s is the cuttings density in ppg, μ_a is the mud apparent viscosity in centipoise, μp is the mud plastic viscosity in centipoise, τ_y is the mud yield value lb/100 ft², d_s is the equivalent spherical diameter of cutting in inches, and V_a is the average annular fluid velocity (fpm).

$$V_a = 60Q/(2.448(D_h^{\ 2} - D^2)\ \text{ft/min}$$

Q is the flow rate in gpm, D_h is the inside diameter of casing or diameter of the hole in inches, and D is the outside diameter of the pipe in inches.

10.02.2 Moore Correlation

The slip velocity based on the Moore correlation is calculated for various conditions from the following equations

$$v_s = 9.24(\rho_s - \rho_m)/\rho_m \quad \text{for } N_r \text{ greater than 2000}$$

$$v_s = 4972(\rho_s - \rho_m)/d_s^{\ 2}/\mu_a \quad \text{for } N_r \text{ less than or equal to 1}$$

N_r is the particle Reynolds number and μ_a is the apparent viscosity. Thus:

$$v_s = [174d_s(\rho_s - \rho_m)^{0.667}]/(\rho_m\mu_a)^{0.333}$$

for 1 is less than or equal to N_r which is less than 2000. Thus:

$$\mu_a = K/144[(D_h - D_{op})/(V_a/60)]^{1-n}[(2 + 1/n)/00208]^n$$

$$K = (510 \times \theta_{300})/511^n$$

In these equations, n is the mud power law index, na = 3.32(log $\theta_{600}/\theta_{300}$), D_h is the hole diameter in inches, and D_{op} is the OD of the pipe in inches.

10.02.3 Walker-Mays Correlation

The slip velocity based on the Walker-Mays correlation is goven for various coditions by the relationship:

$$v_s = 131.5[h_s(\rho_s - \rho_m)/\rho_m]^{1/2}$$

when the Reynolds number is greater than 100

$$\mu_a = 511(\tau/\gamma)$$

$$\tau = 7.9[h_s(\rho_s - \rho_m)]^{1/2}$$

$$v_s = 1.22\tau[d_s\gamma/\rho_m^{1/2}]^{1/2}$$

In these equations, hs is the cutting thickness and γ is the shear rate corresponding to the shear stress, τ.

10.03 Increased Equivalent Circulating Density (ECD) Due to Cuttings

The effective density exerted by a circulating fluid against the formation that considers the pressure drop in the annulus above the point being considered. The equivalent circulating density is an important parameter n avoiding kicks and losses, particularly in wells that have a narrow window between the fracture gradient and the pore-pressure gradient.

The effective mud density (in pounds per gallon) in the hole due to cuttings generation is given by the equation:

$$\rho_{eff} = (\rho_m Q + 141.4296 \times 10^{-4} \times ROP \times d_b^{2})/(Q + 6.7995 \times 10^{-4} \times ROP \times d_b^{2})$$

ρ_m is the mud density without cuttings in ppg, Q is the mud flow rate in gpm, ROP is the rate of penetration, ft/hr, and d_b is the diameter of the bit or the diameter of the hole drilled in inches. The density due to the drill cuttings is obtained from the equation:

$$\rho_{eff} - \rho_f = \rho_c = (ROP \times d_b^{2} \times 10^{-4}(141.4296 - 6.7995 \times \rho_f)/(Q + 7.6995 \times 10^{-4} \times ROP \times d_b^{2})$$

Thus, the equivalent circulating density (ECD) is:

$$ECD = \rho_f + \Delta\rho_a + \Delta\rho_c \text{ ppg}$$

ρ_m is the mud density, ppg, $\Delta\rho_a$ is the equivalent mud weight increase due to annular frictional pressure losses, and $\Delta\rho_c$ is the equivalent mud weight increase due to cuttings.

Example calculation

A 17½-inch hole was drilled at the rate of 120 feet per hour with a circulation rate of 1000 gpm. The mud density was 8.8 ppg. Calculate the effective mud weight due to cuttings.

Solution:

The effective mud density in the hole due to cuttings generation is

$$\rho_{eff} = (\rho_m Q + 141.4296 \times 10^{-4} \times ROP \times d_b^{\,2}) / (Q + 6.7995 \times 10^{-4} \times ROP \times d_b^{\,2})$$

Thus:

$$\rho_{eff} = (8.8 \times 1000 \times 141.4296 \times 10^{-4} \times 120 \times 17.5^2) / (1000 + 6.7995 \times 10^{-4}\, 120 \times 17.5^2) = 9.09 \text{ ppg}$$

10.04 Keyseating

The keyseat is a section of a borehole, usually of abnormal deviation and relatively soft formation, that has been eroded or worn by drill pipe to a size smaller than the tool joints or collars of the drill string; the keyhole-type configuration resists passage of the shoulders of the pipe upset (box) configurations when pulling out of the hole. The side force to create a keyseat is given by the relationship:

$$F_L = T \sin \beta$$

$$F_L = 2T \sin \beta/2$$

$$F_L = T\, \beta L$$

F_L is the lateral force, T is the tension in the drill string just above the keyseat area in lbf, β is the dogleg angle, and L is the length of the dogleg in feet. Thus, the depth of the formation cut due to keyseating is given by the relationship:

$$d_{key} = C_{sc}[T\kappa L_{dl})/OD_{tj}]^{1/2}\,(L_{tj}/ROP)N$$

where d_{key} is the depth of the keyseat, C_{sc} is the side cutting coefficient of the tool joint, T is the drill string tension at the keyseat, κ is the dogleg curvature, L_{dl} is the length of the dogleg, OD_{tj} is the outside diameter of the tool joint, N is the speed in rpm, L_{tj} is the length of the tool joint, and ROP is the rate of penetration. The side cutting coefficient C_{sc} depends on the formation type, the formation stress, the deviation of the hole, and the wellbore pressure.

When the rate of penetration is zero, the above equation reduces to:

$$d_{key} = C_{sc}[T\kappa L_{dl})/OD_{tj}]^{1/2} Nt$$

where t is the total rotating time.

Example calculation

Calculate the side force in the keyseat with the dogleg angle of 4 degrees per 100 feet in a 100 feet keyseat interval. The effective tension above the keyseat is found to be 120 kip.

Solution:

The side force at the keyseat is given as:

$$F_L = T \sin \beta = 120{,}000 \times \sin(4 \times \pi/180) = 8370 \text{ lbf}$$

$$F_L = 2T \sin \beta/2 = 2 \times 120{,}000 \times \sin(4/2 \times \pi/180) = 8376 \text{ lbf}$$

$$F_L = T\beta L = 120{,}000 \times (4/100 \times \pi/180) \times 100 = 8377 \text{ lbf}$$

10.05 Lost Circulation

Lost circulation (circulation loss) refers to a lack of mud returning to the surface after being pumped down a well which occurs when the drill bit encounters natural fissures, fractures or caverns, and mud flows into the newly available space. Lost circulation may also be caused by applying more mud pressure (that is, drilling overbalanced) on the formation than it is strong enough to withstand, thereby opening a fracture which serves as a mud loss channel.

Thus, if $V_l < V_{an/dp}$, the length of the low-density fluid required to balance the formation pressure is given by the equation:

$$L_l = V_l/C_{an/dp}$$

However, if Vl > Van/dc, the length of the low-density fluid required to balance the formation pressure is given by the equation:

$$L_l = L_{dc} + (V_l - V_{an/dp)/} C_{an/dp}$$

V_l is the volume of low-density fluid pumped to balance the formation pressure in bbls, $C_{an/dc}$ is the annulus capacity behind the drill collar in bbl/ft, $C_{an/dc}$ is the annulus capacity behind the drill pipe in bbl/ft, $V_{an/dc}$ is the annulus volume against the drill collar in bbls, $V_{an/dp}$ is the annulus volume against the drill pipe in bbls, and L_{dc} is the length of the drill collar in feet.

The formation pressure is given by:

$$p_{ff} = 0.052 \times D_w \times \rho_w + 0.052 \times (D_v - D_w) \times \rho_m$$

D_v is the vertical depth of the well where the loss occurred in feet, ρ_m is the density of the mud in ppg, D_w is the water depth in feet, and ρ_w is the seawater density in ppg.

Example calculation

While drilling an 8.5-inch hole at 17,523 ft (TVD) with a mud density of 11 ppg, the well encountered a big limestone cavern that resulted in mud loss. Drilling was stopped, and the annulus was filled with 58 bbls of 8.4 ppg water until the well was stabilized.

Calculate the formation pressure and the density that should be used to drill through the zone. The previous 9.625-inch casing was set at 15,500 ft. The drilling consists of 900 ft of 6-inch drill collar and 5-inch drill pipe. Use the capacity of the casing annulus against the drill pipe to be 0.05149 bbl/ft.

Solution:

The volume of the casing annulus against the drill pipe is:

$$V_{an/dp} = 0.05149 \times 15500 = 798 \text{ bbl}$$

and $V_l < V_{an/dp}$

The length of the annulus when balanced is:

$$L_l = 58/0.05149 = 1126.43 \text{ feet}$$

The formation pressure is:

$$\begin{aligned} p_{ff} &= 0.052 \times 1127 \times 8.4 + 0.052 \times (17523 - 1127) \times 11 \\ &= 492.27 + 9378.5 = 9870.8 \text{ psi.} \end{aligned}$$

The equivalent mud weight for drilling = 9870.8/(0.052 × 17523) = 10.83 ppg

10.06 Common Minerals and Metals Encountered During Drilling Operations

Name	Formula	Specific gravity	lb/gal
Aluminum	Al	2.70	22.5
Barite	$BaSO_4$	4.2	35.0
Brass (red)		8.75	72.9
Calcite	$CaCO_3$	2.72	22.7
Steel, Stainless		8.02	66.8
Steel (13 Cr)		7.75	64.6
Steel, Carbon		7.8	65
Chromium	Cr	7.19	59.9
Copper	Cu	8.96	74.6
Diatomaceous Earth		0.4–0.6	3.3–5.0
Dolomite	$CaMg(CO_3)_2$	2.85	23.7
Feldspar		2.57–2.76	21.4–23.0
Galena	PbS	7.5	62.5
Gypsum	$CaSO_4+2H_2O$	2.32	19.2
Halite (salt)	NaCl	2.16	18.0
Hematite	Fe_2O_3	4.9–5.3	40.8–44.1
Iron	Fe	7.86	65.5
Montmorillonite (bentonite)		2.5	20.8
Quartz	SiO_2	2.65	22.1
Sepiolite (clay)		2	16.7
Siderite	$FeCO_3$	3.9	32.5
Zinc	Zn	7.14	59.5

10.07 Mud Weight Increase Due to Cuttings

The volume of the cuttings entering the mud system is derived from the equation:

$$V_c = [1 - \varphi)D_b^2 \times ROP)/1029 \text{ bbl/hr}$$

where φ is the average formation porosity, Db is the diameter of the bit in inches, and ROP is the rate of penetration in feet per hour.

Or:

$$V_c = [1 - \varphi)D_b^{\,2} \times ROP)/24.49 \text{ gallons per hour}$$

$$V_c = [1 - \varphi)D_b^{\,2} \times ROP)/1469.4 \text{ gallons per minute}$$

An estimate of the average annulus mud weight ($\rho_{m,av}$) can be formulated from the above equations and given by:

$$\rho_{m,av} = (\rho_{ps}Q + 0.85D_h^{\,2}ROP)/(Q + 0.0408D_h^{\,2}ROP)$$

where $\rho_{m,av}$ is the average annular mud weight in lb/gal, Q is the flow rate in gpm), ρ_{ps} is the measured mud weight at the pump suction in lb/gal, D_h is the diameter of hole in inches, and ROP is the penetration rate based on the time the pump is on before, during, and after the joint is drilled down (f_{pm}).

Example calculation
Calculate the volume of the cuttings generated while drilling a 12.25-inch hole with the rate of penetration of 50 ft/hr. Assume that the formation porosity is 30%.

Solution:
The volume of the cuttings generated in barrels per hour is:

$$V_c = [1 - \varphi)D_b^{\,2} \times ROP)/1029 = [(1 - 30/100) \times 12.5^2 \times 50]/1029 = 5.1 \text{ bbl/hr}$$

$$V_c = [1 - \varphi)D_b^{\,2} \times ROP)/24.49 = [(1 - 30/100) \times 12.5^2 \times 50]/24.49 = 214.5 \text{ gph}$$

$$V_c = [1 - \varphi)D_b^{\,2} \times ROP)/1469.4 = [(1 - 30/100) \times 12.5^2 \times 50]/1469.4 = 3.57 \text{ gpm}$$

10.08 Pressure Loss in the Drill String

All pressure losses, at first, assume a laminar flow regime. Power Law Model calculations begin with:

$$P_{lf} = [(L \times k)/(300 \times d)] \times [1.6 \times V_p)/d \times (3n + 1)/4n]^n$$

P_{lf} is the pressure loss in laminar flow in psi, L is the length of section in feet, V_p is the velocity in the section of drill string in ft/min, *d* is the inside Diameter of drill string in inches, k is the consistency index, and *n* is the power index. The fluid velocity in the drillstring can be determined by using:

$$V_p = 24.51 \times Q)/d_1^2$$

Q is the pump flow rate in gpm, and d_1 is the pipe ID in inches. The equivalent viscosity (m) is then determined, using:

$$\mu = (90000 \times P_{lf} \times d^2)/(L \times V)$$

Which, in turn, is used to determine the Reynolds Number.

$$Re = (15.46 \times MD \times V_p \times d)/\mu$$

The flow behavior, as calculated using the with the power law model, will vary depending on the "n" value of the fluid.

10.09 Spotting Fluid Requirements

Spotting fluid is the term used for a small volume or pill of fluid placed in a wellbore annulus to free differentially stuck pipe. Oil-base drilling mud is the traditional stuck-pipe spotting fluid. Speed in mixing and placing the spot is of primary importance to successfully freeing pipe.

The height of the spotting fluid in the annulus is calculated from:

$$L_s = \Delta p/[0.052(\rho_m - \rho_s)] \text{ feet}$$

Δp is the differential pressure due to spotting fluid in psi, ρ_m is the mud weight in ppg, and ρ_s is the spotting fluid density, ppg.

Example calculation

Estimate the height required of an un-weighted soak solution of 7.5 ppg to reduce the hydrostatic pressure of 200 psi at the free point. The mud weight in the hole is 9.5 ppg.

Solution:

The height of the spotting fluid in the annulus is calculated as:

$$L_s = \Delta p/[0.052(\rho_m - \rho_s)] = 200/[0.052(9.5 - 7.5)] = 1442 \text{ feet}$$

11

Cementing

11.01 API Classification of Cement

API Class	Application
A	– Used at depth ranges of 0 to 6000 ft. – Used at temperatures up to 170 °F – Used when special properties are not required – Used when well conditions permit – Economical when compared to premium cements
Neat	* Normal Sluny Weight is 15.6 ppg * Normal Mixmg Water Requirement – 46% (5.19 gal/sk & 0.693 ft^3/sk) * Normal Slurry Yield – 1.17 ft^3/sk
B	– Used at depth ranges of 0 to 6000 ft. – Used at temperatures up to 170 °F – Used when moderate to high sulfate resistance is required – Used when well conditions permit – Economical when compared to premium cements

(*Continued*)

API Class	Application
Neat	* Normal Slurry Weight is 15.6 ppg * Normal Mixmg Water Requirement – 46% (5.19 gal/sk & 0.693 ft^3/sk) * Normal Slurry Yield – 1.17 ft^3/sk
C	– Used at depth ranges of 0 to 6000 ft. – Used at temperatures up to 170 °F – Used when lngh early-streugtli is required – Used when its special properties are required – High m tricalcium silicate
Neat	* Normal Shiny Weight is 14.8 ppg * Normal Mixing Water Requirement – 56% (6.31 gal/sk & 0.844 ft^3/sk) * Normal Slurry Yield – 1.32 ft^3/sk
D/E	– Class D used at depths from 6000 to 10000 ft. and at temperatures from 170 – 60 °F – Class E used at depth from 10000 to 14000 ft. and at temperatures from 170 – 290 °F – Both used at moderately high temperatures and high pressures – Both available m types that exhibit regular and high resistance to sulfate
Neat	– Both are retarded with an organic compound, chemical composition and grind – Both are more expensive than Portland cement * Normal Slurry' Weight – 16.5 ppg * Normal Mixmg Water Requirement – 38% (4.28 gal/sk & 0.572 ft^3/sk) * Normal Slurry' Yield – 4.29 ft^3/sk
F	– Used at depth ranges of 10000 to 16000 ft. – Used at temperatures from 230 – 320°F – Used when extremely high temperatures and pressures are encountered – Available in types that exhibit moderate and high resistance to sulfate – Retarded with an organic additive, chemical composition and grind
Neat	* Normal Sluny' Weight – 16.5 ppg * Normal Mixing Water Requirement – 38% (4.28 gal/sk & 0.572 ft^3/sk) * Normal Sluny – Yield – 1.05 ft^3/sk

(*Continued*)

API Class	Application
G,H	- Used at depth ranges from 0 to 8000 ft. - Used at temperatures up to 200°F without modifiers - A basic cement compatible with accelerators or retarders - Usable over the complete ranges of A to E with additives - Additives can be blended at bulk station or at well site - Class H is a coarser grind than Class G
Neat	* Class G sluny' weight is 15.8 ppg * Class G mixing water requirement – 44% (4.96 gal/sk & 0.663 ft³/sk) * Class G slurry yield – 1.14 ft³/sk * Class H slurry' weight is 15.6 (shallow) to 16.4 (deep) ppg * Class H API water requirement – 46% (5.2 gal sk & 0.69ft³/sk) to 38% (4.3 gal sk & 0.57 ft³/sk) * Class H slimy' yield – 1.17 ft³/sk (shallow) to 1.05 ft³/sk (deep)
J	- Used at depth ranges from 12000 to 16000 ft - Used for conditions of extreme temperature and pressure: 170–320°F (unmodified) - Usable with accelerators and retarders - Will not set at temperatures less than 150°F when used as a neat slurry * Water requirements set by manufacturer

11.02 Cement: Physical Properties of Additives

Material	Bilk W eight lbs/ft³	Specific gravity glee	Absolute volume gal. lb ft³ Lb	
Sodium Chloride	71.0	2.17	0.0553	0.0074
Calcium Chloride	56.0	1 96	0.0612	0.0082
Potassium Chloride	64.9	1.984	0.0604	0.0081
Gypsum	75	2.7	0.0444	0.0059
Cement	94	3.14	0.0382	0.0051
Attapulgite	40	2.89	0.0415	0.0053
Barite	135	4.23	0.0284	0.0038
Hematite	193	5.02	0.0239	0.0032

Material	Bilk W eight Ibs/ft³	Specific Gravity glee	Absolute Volume gal. lb ft³ Lb	
Diatomaceous Earth	16.7	2.1	0.0572	0.0076
Pozzolan	40	2.43	0.0493	0.0066
Diesel Oil (1)	51.1	0.82	0.1457	0.0195
Diesel Oil (2)	53.0	0.85	0.1411	0.0189
Fly Ash	74	2.46	0.0487	0.0065
Bentonite	60	2.65	0.0453	0.0060
Gilsomte	50	1.07	0.1122	0.0150
Nut Plug	48	1.28	0.0938	0.0125
Silica Flour	70	2.63	0.0456	0.0061
Sand	100	2.63	0.0456	0.0061
Water (fresh)	624	1.00	0.1200	0.0160
Water (Sea)	63.96	1.025	0.1169	0.0153
Lignosulphonate	35.1	1.36	0.0882	0.0118
Polymer (FL-50)	35	1.34	0.0895	0.0119

11.03 Cement Plug

The cement plug is part of the well completion process in which a plug of cement slurry placed in the wellbore. Cement plugs are also used for a variety of applications including hydraulic isolation, provision of a secure platform, and in window-milling operations for sidetracking a new wellbore.

The number of sacks of cement required for placing a cement plug can be estimated from the equation:

$$N_c = (L_p \times V_h)/Y$$

L_p = is the length of plug in feet ft, Y is the yield in cubic feet per sack, and V_h is the capacity of the hole in cubic feet per foot. The spacer volume ahead of the slurry is calculated from the equation:

$$V_{sa} = (V_{sb} \times C_{an})/V_{db}$$

V_{sb} is the volume of spacer behind the slurry, C_{an} is the annular volume in feet, and V_{dp} is the pipe volume in cubic feet. The length of the plug can be calculated from the equation:

$$L_p = (N_c \times Y)/V_{dp} + V_{an}) \text{ feet}$$

The volume of mud required to displace the pipe can be calculated from the equation:

$$Vm = (L_{dp} = L_p)V_{dp} - V_{sb} \text{ bbls}$$

L_{dp} is the length of drill pipe in feet.

Example calculation

A cementing engineer is planning to spot a cement plug of 200 ft with a top of the plug at 5000 ft. The hole size = 8.5 inches. He assumes an excess factor of 20%. The pipe used for spotting is 5 inches with an internal diameter of 4.276 inches. A slurry yield of 1.2 cubic feet/sack is used. What would be the number of sacks of cement needed for this balanced cement plug job? Calculate the length of the plug before the pipe is pulled out.

Solution:

The length of the plug desired is (5200 – 5000 ft =) 200 ft. The slurry volume desired is: $(8.5^2 \times 200)/1029.5 = 16.85$ bbls. Therefore, the number of sacks of cement needed is: $(5.615 \times V_{cem})/Y = (5.615 \times 16.85)/1.2 = 79$ sacks

The hole capacity is: $8.5^2/1029.5 \times 1.2 = 0.084216$ bbls/ft. Therefore, the annulus capacity is: $(8.^{52} - 5^2)/1029.5 = 0.04590$ bb/ft. The pipe capacity is: $4.276^2/1029.5 = 0.01776$ bbl/ft.

The plug length before the pipe is pulled out is calculated from the equation:

$$\begin{aligned} L_{plug(before)} &= L_{plug(after)} \times \text{hole capacity})/(\text{annulus capacity} \\ &\quad + \text{pipe capacity}) \\ &= (200 \times 08421)/(0.04590 + 0.01776) = 265 \text{ feet} \end{aligned}$$

Or:

$$\begin{aligned} L_{plug(before)} &= \text{Number} \times \text{yield})/(\text{annulus capacity} + \text{pipe capacity}) \\ &= (79 \times 1.2)/5.615(0.04590 + 0.01776) = 265 \text{ feet} \end{aligned}$$

The hole capacity is calculated as (8.52)

$$\frac{(8.5^2)}{1029.5} \times 1.2 = 0.084216 \text{ bbl/ft.}$$

The annulus capacity is

$$\frac{\left(8.5^2 - 5^2\right)}{1029.5} = 0.04590 \text{ bbl/ft.}$$

The pipe capacity is

$$\frac{\left(4.276^2\right)}{1029.5} = 0.01776 \text{ bbl/ft.}$$

The plug length before the pipe is pulled out is given by

$$L_{Plug\,(before)} = \frac{L_{Plug\,(after)} \times \text{ hole capacity}}{(\text{annulus capacity} + \text{ pipe capacity})},$$

$$= \frac{200 \times 0.08421}{(0.04590 + 0.01776)} = 265 \text{ ft.}$$

Alternatively, it can be calculated as

$$L_{Plug\,(before)} = \frac{\text{Number of sacks} \times \text{ Yield}}{(\text{annulus capacity} + \text{ pipe capacity})},$$

$$= \frac{79 \times 1.2}{5.615 \times (0.04590 + 0.01776)} = 265 \text{ ft.}$$

11.04 Cement Slurry Requirements

Cement is used to hold casing in place and to prevent fluid migration between subsurface formations and it is necessary to know the numbers of sacks of cement required. The number of sacks of cement, N_c, is calculated from the equation:

$$N_c = V_{sl}/Y \text{ sacks}$$

V_{sl}/ is the volume of slurry in cubic feet and Y is the yield of cement in cubic feet per sack. As a general guide, the sack, which is a unit of measure for Portland cement in the United States, refers to the amount of cement that occupies a bulk volume of 1.0 ft^3. For most Portland cement, including API classes of cement, a sack weighs 94 pounds.

11.05 Contact Time

The contact time is the elapsed time required for a specific fluid to pass a designated depth or point in the annulus during pumping operations. Contact time is normally used as a design criterion for mud removal in turbulent flow.

An estimation of the volume of cement needed for the removal of mud cake by turbulent flow is given by the equation:

$$V_t = t_c \times q \times 5.616 \text{ cubic feet per barrel}$$

Example calculation
The water requirement for API class G cement is 5.0 gal/94-lb sack, or 44%. Calculate the density of the slurry, water requirement, and slurry yield. Assume the specific gravity of cement = 3.14.

Solution:

Water needed for 100 g cement = 100 × 0.44 = 44 gallons
The volume of cement, V_c, is equal to: 100/3.14 = 31.85 cm^3
The volume of water, V_w, is equal to: = 44 cm^3.
Total slurry volume = 75.85 cm^3
Total mass of slurry = 100 + 44 = 144 g
The slurry density, ρ_{sl}, = $(W_c + W_w + W_a)/(V_c + V_w + V_a)$
= 144/75.85 = 1.8948 = 18.8 ppg.

Therefore, the water requirement per sack is: The water requirement per sack is:

= 426.38 × 44 = 18760 cm^3,
= 18760/3785.4 = 4.95 gallons per sack.

The slurry yield is:

69.85 cm^3/100 g = 0.6985 cm^3/g = 1.140 ft^3/sack

11.06 Gas Migration Potential

Gas migration, in the current context, is the gas entry into a cemented annulus creating channels with the potential to provide a flow path of formation fluids, including hydrocarbons, into the wellbore. This can cause gas/fluid flow in the annulus and can usually be detected by cement bond logs and/or by noticing unwanted pressures.

The pressure reduction for the cement column is given by

$$Pr = 1.67 \times L/(D_h - D_p)$$

Thus, the gas migration potential, GMP, is: $P_{r,max}/P_{ob}$ where Dh is the diameter of the hole in inches, D_p is the outside diameter of the casing pipe in inches, P_{ob} is the reservoir pressure in psi, and L is the length of the pipe column exposed to the cement from the reservoir zone in feet. Ranges for the gas migration potential are: (i) 0 to 3 = low, (ii) 3 to 8 = moderate, and (iii) >8 = high.

Example calculation
Calculate the gas migration potential using the following data:

- Depth: 10000 ft, 8 ½-inch hole
- Casing OD = 7 inches
- Mud density = 16 ppg
- Active formation depth = 9000 feet
- Reservoir pressure = 7000 psi
- Cemented depth top = 7500 feet
- Slurry density = 17.5 ppg

Solution
Hydrostatic pressure = 7500 × 0.052 × 16 + 1500 × 0.052 × 17.5 = 7605 psi.

$$P_{ob} = 7605 - 7000 = 605 \text{ psi.}$$

$$P_{r,max} = 1.67 \times 1500/(8.5 - 7) = 1670 \text{ psi.}$$

$$GMP = 1670/605 = 2.76 \text{ (low)}$$

11.07 Hydrostatic Pressure Reduction

The hydrostatic pressure is the pressure exerted by a fluid at equilibrium at a given point within the fluid, due to the force of gravity. The hydrostatic pressure increases in proportion to depth measured from the surface because of the increasing weight of fluid exerting downward force from above. Hydrostatic pressure reduction due to the spacer is:

$$\Delta p = 0.292(\rho_m - \rho_s)C_{an} \times V_s \text{ psi})$$

ρ_m is the density of the mud in ppg, ρ_s is the density of the spacer in ppg, C_{an} is the annular capacity in ft^3/ft, and Vs is the spacer volume in bbls.

11.08 Portland Cement – Typical Components

Tricalcium Silicate – $3Ca0:Si0_2$ – C_3S
1. Tlie major compound in Portland cement 2. Contributes to strength development, especially during the fust 28 days 3. Hydration equation: $2(3Ca0:Si0_2) + 6H_20 \text{ ---> } 3Ca0:2Si0_2:3H_20 + 3Ca(OH)_2$
Dicalcium Silicate – $2CaO:Si0_2$ – C_2S
1. A much slower hydrating compound than tricalcium silicate 2. Contributes to a slower, gradual increase in strength, over an extended period of time 3. Hydration equation: $2(2Ca0:Si0_2) + 4H_20 \text{ ---> } 3Ca0:2Si0_2:3H_20 + Ca(OH)_2$
Tricalcium Aluminate – $3CaO:Al_20_3$ – C_3A
1. Promotes rapid hydration of cement 2. Controls the setting time of cement 3. Regulates the cements resistance to sulfate attack (HSR – High Sulfate Resistance) 4. Gypsum usually added to control hydration and flash setting 5. Produces most of the heat observed over first few days 6. Hydration equation: $3Ca0:Al_20_3 + 12H,0 + Ca(OH)_2 \text{ ---> } 3Ca0:Al_20_3:Ca(0H)_2:12H_20$ $3Ca0:Al_20_3 + 10H,O + CaS0_4:2H_20 \text{ ---> } 3Ca0:Al_20_3:CaS0_4:12H_20$
Tetracalcium Alummofenite – $4Ca0:Al_20_3:Fe_20_3$ – C_4AF
1. Promotes low heat of hydration in cement 2. Hydration equation: $4Ca0:Al_20_3:Fe_20_3 + 10H_2O + 2Ca(OH)_2 \text{ ---> } 6Ca0:Al_20_3:Fe_20_3:12H_20$

11.09 Slurry Density

The slurry density is the weight per unit volume of a cement slurry and is usually given in or lbm/gallon. A typical oil-well or gas-well slurry has a density on the order of 11.5 lbs/gallon to 19.0 lbs/gallon, although special techniques, such as foamed cementing and particle-size distribution cementing, extend this range to 7 lbs/gallon to 23 lbs/gallon.

The slurry density is calculated form the equation:

$$\rho_{sl} = (W_c + W_w + W_a)/(V_c + V_w + V_a)$$

(W_c is the weight of the cement in lbs, W_w is the weight of the water in lbs, and W_a is the weight of the additive in lbs. V_c, V_w, and V_a are the respective volumes of the cement, water, and additive.

11.10 Yield of Cement

The yield of cement is the volume occupied by one sack of dry cement after mixing with water and additives to form a slurry having the desired density. The yield is commonly expressed (in the United States) units as cubic feet per sack (cubic feet per sack, ft³/sk). Thus:

$$Y_{sl} = V_{sl}/7.48 \text{ ft}_3\text{/sk}$$

Ysl is the yield of slurry, V_{sl} – the volume of the slurry, is equal to $V_c + V_w + V_a$ gallons in which V_c is the volume of cement, V_w is the volume of water, and V_a is the volume of additive.

The mix water requirement is:

$$V_{mw} = Vms_i + N_c \text{ cubic feet}$$

where Vms_i is the mix water per sack.

The number of sacks of the additive is:

$$N_a = N_c \times A_{\%}$$

where A% is the percentage of additive.

Based on a US sack containing 94 lbs, the weight of the additive is:

$$W_a = N_a \times 94 \text{ lbs}$$

12

Well Cost

12.01 Drilling Cost

The drilling cost is a summation of several variable costs that vary with the job. The formula and example given below are merely an examples of how part of the costs should be calculated, assuring that all aspects of the cost per item are included. Thus, the overall well cost excluding the production is calculated as:

$$C_{wdo} = C_d + C_o$$

C_d is the cost of drilling a foot of hole that includes bit cost, downhole drilling tools cost, and rig rental costs only, C_o is all other costs of making a foot of hole, including the costs of casings, mud, cementing services, logging services, coring services, site preparation, fuel, transportation, and completion. The bit run costs may be expressed as

$$C_{di} = [C_{bi} + C_r(T_{di} + T_{ti} + T_{ci})]/\Delta D_i$$

C_{di} is the drilling cost in dollars per foot, C_{bi} is the bit cost in dollars, C_r is the rig cost in dollars per hour, T_{di} is drilling time in hours, T_{ti} is the trip time in hours, T_{ci} is the connection time, and ΔD_i is the formation interval drilled in feet, by bit number *i*. The trip time is given as:

$$T_{ti} = 2(t_s/L_s)/D_i$$

T_{ti} = the trip time required to change a bit and resume drilling operations in hours, t_s is the average time required to handle one stand of drill string in hours, L_s is the average length of one stand of drill string in feet, and D_i is the depth where the trip was made in feet. Assuming that the average bit life for a given bit group in a given hole interval is T_b the depth of the next trip is given by the following equation:

12.02 Expected Value

The expected value is the difference between expected profits and expected costs. Expected profit is the probability of receiving a certain profit times the profit, and the expected cost is the probability that a certain cost will be incurred.

$$\text{Expected value, EV} = \Sigma p_i C_i$$

In this equation, p_i is the probability of the ith event and C_i is the cost of the ith event.

12.03 Future Value

The future value (FV) is the value of a current asset at a specified date in the future based on an assumed rate of growth over time.

$$FV = PV(1 + r/n)^{n \times m}$$

PV is the present value, *r* is the periodic interest rate or growth rate in fraction, *n* is the number of payments per year, and *m* is the number of years. For continuous compounding, the equation is:

$$FV = PVe^{r \times n}$$

12.04 Price Elasticity

Price elasticity is the ratio of the percentage of change of wells and footage drilled to the percentage change in the crude price and describes the degree of responsiveness of the rig in demand or rig in supply to the change in the crude price. Price elasticity is used to measure the effect of economic variables such as demand or supply of rigs or wells drilled with respect to change in the crude oil price and enables the company to discover the sensitivity one variable is with the other one.

Price elasticity € is equal to the percentage change in drilling wells divided by the percentage change in crude oil prices. Thus:

$$E = (\%\Delta R)/(\%\Delta P)$$

13

Appendices

Acronyms in Common Use in the Petroleum and Natural Gas Industries

Acronym	*Description*
AC	Alternating Current
AGA	American Gas Association
API	American Petroleum Institute
CCR	Central Control Room
CMS	Condition Monitoring Systems
CSP	Collector and Separation Platform
DC	Direct Current
DYNPOS	Dynamic positioning (of rigs and ships)
E&P	Exploration and Production
EOR	Enhanced Oil Recovery
ESD	Emergency Shutdown system
ESP	Electric Submerged Pump
F&G	Fire & Gas System
FPSO	Floating Production Storage and Offloading
GB(S)	Gravity Base Structure

GOR	Gas Oil Ratio from the well
GOSP	Gas Oil Separation Plant
GTP	Gas Treatment Platform
HP	High Pressure
HPU	Hydraulic Power Unit (topside utility for subsea)
HVAC	Heat Ventilation and Air Conditioning
IR	Infrared
ISO	International Standards Organization
K-Mass Flow	Coriolis type Mass Flow meter
LNG	Liquid Natural Gas (e.g. Methane)
LP	Low Pressure
LPG	Liquefied Petroleum Gas (e.g. Propane)
MCC	Motor Control Centre
MTBF	Mean Time Between Failure
NGL	Natural Gas Liquids, Condensates see also LPG
PCP	Progressive Cavity Pump
PD-Meter	Positive Displacement meter
PGP	Power Generation Platform
PID	Proportional Integral Derivate control algorithm
PIMS	Production Information Management System
PoC	Pump of controller (for artificial lift)
POSMOOR	Position mooring for a floating facility
PSD	Process Shutdown System
ROV	Remote Operated Vehicle (for subsea workover)
RTU	Remote Terminal Unit
SAS	Safety and Automation System
SCADA	Supervisory Control and Data Acquisition
TIP	Tie-In Platform
TLP	Tension Leg Platform
UMS	Unmanned Machinery Space classification (marine = E0)
URF	Umbilicals, Risers, and Flowlines
UV	Ultraviolet
WHP	Wellhead Platform

Table Compositions of Selected Common Alloys*.

Alloy class	Example	Constituents									
		Ni	Cr	Mo	Fe	Co	Ti	Cu	Cb	A1	V
C arbon Steel	CIO				> 94						
Low-Alloy Steel	l-l/4Cr l/2Mo		1.25	0.5	balance						
Fe-Ni-Cr + Mo	Type 316L	13.0	17.0	2.3	balance						
	Alloy SO OH	32.5	21 0		4.6						
	20C:b-3	35.0	20 0	2.5	balance			3.5			
Ni-Cr-Mo	Alloy C2	54.0	15.5	16.0							
	Alloy C276	57.0	16.0	16.0	5.5						
	Alloy C4	54.0	16.0	15.5	3.0						
	Alloy 625	60.0	21 5	9.0					3.7		
Ni-Cr-Fe	Alloy G	450	22.2	6.5	19.5			2.0			
	Alloy 600	76.0	15.0		8.0						
Ni-Mo	Alloy B2	balance	1.0	28.0	2.0	10					
NLCu	Alloy 400	65 1						32.0			
Nickel	Alloy 200	99.9									
Co-Base	ULTIMET (R)	90	26.0	5.0	3.0	54.0					
Ti-Base	Ti-6Al-4V						90			6.0	4.0

*Alloys that are of use in the petroleum and natural gas industries.

Casing and Liner Capacity and Displacement

Nominal size inches	Weight w/ coupling lb/ft	OD inches	ID inches	Capacity bbl/ft	Capacity ft/bbl	Pipe displacement bbl/100 ft
4-1/2	9.50	4.500	4.090	0.0162	61.54	0.342
4-1/2	10.50	4.500	4.052	0.0159	62.70	0.372
4-1/2	11.60	4.500	4.000	0.0155	64.34	0.413
4-1/2	13.50	4.500	3.920	0.0149	67.00	0.474
4-1/2	15.10	4.500	3.826	0.0142	70.33	0.545
5	11.50	5.000	4.560	0.0202	49.51	0.409
5	13.00	5.000	4.494	0.0196	50.97	0.467
5	15.00	5.000	4.408	0.0189	52.98	0.541
5	18.00	5.000	4.276	0.0178	56.30	0.652
5	21.40	5.000	4.126	0.0165	60.47	0.775
5	23.20	5.000	4.044	0.0159	62.95	0.840
5	24.10	5.000	4.000	0.0155	64.34	0.874
5-1/2	14.00	5.500	5.012	0.0244	40.98	0.498
5-1/2	15.50	5.500	4.950	0.0238	42.02	0.558
5-1/2	17.00	5.500	4.892	0.0232	43.02	0.614
5-1/2	20.00	5.500	4.778	0.0222	45.09	0.721
5-1/2	23.00	5.500	4.670	0.0212	47.20	0.820
5-1/2	26.80	5.500	4.500	0.0197	50.84	0.971
5-1/2	29.70	5.500	4.376	0.0186	53.76	1.078
5-1/2	32.60	5.500	4.250	0.0175	57.00	1.184
5-1/2	35.30	5.500	4.126	0.0165	60.47	1.285
5-1/2	38.00	5.500	4.000	0.0155	64.34	1.384
5-1/2	40.50	5.500	3.876	0.0146	68.53	1.479
5-1/2	43.10	5.500	3.750	0.0137	73.21	1.572
6-5/8	20.00	6.625	6.049	0.0355	28.14	0.550
6-5/8	24.00	6.625	5.921	0.0341	29.37	0.699
6-5/8	28.00	6.625	5.791	0.0326	30.70	0.846
6-5/8	32.00	6.625	5.675	0.0313	31.97	0.976
7	17.00	7.000	6.538	0.0415	24.08	0.608
7	20.00	7.000	6.456	0.0405	24.70	0.711
7	23.00	7.000	6.366	0.0394	25.40	0.823

(*Continued*)

Casing and Liner Capacity and Displacement (*Continued*)

Nominal size inches	Weight w/ coupling lb/ft	OD inches	ID inches	Capacity bbl/ft	Capacity ft/bbl	Pipe Displacement bbl/100 ft
7	26.00	7.000	6.276	0.0383	26.14	0.934
7	29.00	7.000	6.184	0.0371	26.92	1.045
7	32.00	7.000	6.094	0.0361	27.72	1.152
7	35.00	7.000	6.004	0.0350	28.56	1.258
7	38.00	7.000	5.920	0.0340	29.37	1.355
7	42.70	7.000	5.750	0.0321	31.14	1.548
7	46.40	7.000	5.625	0.0307	32.54	1.686
7	50.10	7.000	5.500	0.0294	34.03	1.821
7	53.60	7.000	5.376	0.0281	35.62	1.952
7	57.10	7.000	5.250	0.0268	37.35	2.082
7-5/8	24.00	7.625	7.025	0.0479	20.86	0.854
7-5/8	26.40	7.625	6.969	0.0472	21.20	0.930
7-5/8	29.70	7.625	6.875	0.0459	21.78	1.056
7-5/8	33.70	7.625	6.765	0.0445	22.49	1.202
7-5/8	39.00	7.625	6.625	0.0426	23.46	1.384
7-5/8	42.80	7.625	6.501	0.0411	24.36	1.542
7-5/8	45.30	7.625	6.435	0.0402	24.86	1.625
7-5/8	47.10	7.625	6.375	0.0395	25.33	1.700
7-5/8	51.20	7.625	6.251	0.0380	26.35	1.852
7-5/8	55.30	7.625	6.125	0.0364	27.44	2.003
7-3/4	46.10	7.750	6.560	0.0418	23.92	1.467
8-5/8	24.00	8.625	8.097	0.0637	15.70	0.858
8-5/8	28.00	8.625	8.017	0.0624	16.02	0.983
8-5/8	32.00	8.625	7.921	0.0609	16.41	1.131
8-5/8	36.00	8.625	7.825	0.0595	16.81	1.278
8-5/8	40.00	8.625	7.725	0.0580	17.25	1.429
8-5/8	44.00	8.625	7.625	0.0565	17.71	1.578
8-5/8	49.00	8.625	7.511	0.0548	18.25	1.746
9-5/8	32.30	9.625	9.001	0.0787	12.71	1.129
9-5/8	36.00	9.625	8.921	0.0773	12.94	1.268
9-5/8	40.00	9.625	8.835	0.0758	13.19	1.417

(*Continued*)

Casing and Liner Capacity and Displacement (*Continued*)

Nominal size inches	Weight w/ coupling lb/ft	OD inches	ID inches	Capacity bbl/ft	Capacity ft/bbl	Pipe Displacement bbl/100 ft
9-5/8	43.50	9.625	8.755	0.0745	13.43	1.553
9-5/8	47.00	9.625	8.681	0.0732	13.66	1.679
9-5/8	53.50	9.625	8.535	0.0708	14.13	1.923
9-5/8	58.40	9.625	8.435	0.0691	14.47	2.088
9-5/8	59.40	9.625	8.407	0.0687	14.57	2.133
9-5/8	64.90	9.625	8.281	0.0666	15.01	2.338
9-5/8	70.30	9.625	8.157	0.0646	15.47	2.536
9-5/8	75.60	9.625	8.031	0.0626	15.96	2.734
10-3/4	32.75	10.750	10.192	0.1009	9.91	1.135
10-3/4	40.50	10.750	10.050	0.0981	10.19	1.414
10-3/4	45.50	10.750	9.950	0.0962	10.40	1.609
10-3/4	51.00	10.750	9.850	0.0942	10.61	1.801
10-3/4	55.50	10.750	9.760	0.0925	10.81	1.972
10-3/4	60.70	10.750	9.660	0.0906	11.03	2.161
10-3/4	65.70	10.750	9.560	0.0888	11.26	2.348
10-3/4	73.20	10.750	9.406	0.0859	11.64	2.631
10-3/4	79.20	10.750	9.282	0.0837	11.95	2.856
10-3/4	85.30	10.750	9.156	0.0814	12.28	3.082
11-3/4	42.00	11.750	11.084	0.1193	8.38	1.477
11-3/4	47.00	11.750	11.000	0.1175	8.51	1.657
11-3/4	54.00	11.750	10.880	0.1150	8.70	1.912
11-3/4	60.00	11.750	10.772	0.1127	8.87	2.140
11-3/4	65.00	11.750	10.682	0.1108	9.02	2.327
11-3/4	71.00	11.750	10.586	0.1089	9.19	2.525
13-3/8	48.00	13.375	12.715	0.1570	6.37	1.673
13-3/8	54.50	13.375	12.615	0.1546	6.47	1.919
13-3/8	61.00	13.375	12.515	0.1521	6.57	2.163
13-3/8	68.00	13.375	12.415	0.1497	6.68	2.405
13-3/8	72.00	13.375	12.347	0.1481	6.75	2.568
16	65.00	16.000	15.250	0.2259	4.43	2.277
16	75.00	16.000	15.124	0.2222	4.50	2.648

(*Continued*)

Casing and Liner Capacity and Displacement (*Continued*)

Nominal size inches	Weight w/ coupling lb/ft	OD inches	ID inches	Capacity bbl/ft	Capacity ft/bbl	Pipe Displacement bbl/100 ft
16	84.00	16.000	15.010	0.2188	4.57	2.982
16	109.00	16.000	14.688	0.2096	4.77	3.911
18-5/8	87.50	18.625	17.755	0.3062	3.27	3.074
20	94.00	20.000	19.124	0.3553	2.81	3.329
20	106.50	20.000	19.000	0.3507	2.85	3.788
20	133.00	20.000	18.730	0.3408	2.93	4.778

Common Conversion Factors

Multiply	by	to Get
acres, ac	0.4047	hectares, ha
acres, ac	43,560	square feet, ft^2
acres, ac	4047	square meters, m^2
acre-feet	43560	cubic feet, ft^3
atmospheres, atm	14.70	pounds per square inch, lb/in^2
bars	0.9869	atmospheres, atm
bars	14.5	pounds per square inch, lb/in^2
barrels, bbl	5.614	cubic feet, ft^3
barrels, bbl	0.159	cubic meters, m^3
barrels, bbl	42	gallons, gal
British thermal units, Btu	252	calories (gram), g-cal
British thermal units, Btu	777.65	foot pounds, ft-lb
Btu/hr	0.29287	watts, W
calories (gram), g-cal	0.003969	British thermal units, Btu
calories (gram), g-cal	4.183	joules, j
Celsius, (degrees)	$(C \times 9/5) + 32$	Fahrenheit (degrees)
centipoise, cp	0.001	newton-sec per sq meter, $N\text{-}sec/m^2$

(*Continued*)

Common Conversion Factors (*Continued*)

Multiply	by	to Get
centipoise, cp	0.002089	pound-sec per sq foot, lb-sec/ft^2
cubic feet, ft^3	0.0370	cubic yards, yd^3
cubic feet, ft^3	0.0283	cubic meters, m^3
cubic feet, ft^3	7.4805	gallons, gal
cubic feet, ft^3	28.32	liters, l
cubic feet per minute, ft/min	0.4719	liters per second, l/sec
Fahrenheit (degrees)	(F-32) × 5/9	Celsius (degrees)
feet, ft	12	inches, in
feet, ft	0.3048	meters, m
feet, ft	30.48	centimeters, cm
feet of water, (head)	0.0295	atmospheres, atm
feet per minute, ft/min	0.5080	centimeters per second, cm/sec
feet per minute, ft/min	0.0183	kilometers per hour, km/hr
feet per minute, ft/min	0.0114	miles per hour, mi/hr
feet per second, ft/sec	0.6818	miles per hour, mi/hr
feet per minute, ft/min	0.3048	meters per minute, m/min
foot pounds, ft-lb	0.001286	British thermal units, Btu
foot pounds, ft-lb	0.3236	calories (gram), g-cal
foot pounds, ft-lb	1.356	newton meters, Nm
gallons, gal (U.S.)	3785	cubic centimeters, cm^3
gallons, gal (U.S.)	0.1337	cubic feet, ft^3
gallons, gal (U.S.)	231	cubic inches, in^3
gallons, gal (U.S.)	0.003785	cubic meters, m^3
gallons, gal (U.S.)	3.7854	liters, l
gallons per minute, gal/min (gpm)	0.0238	barrels per minute, bbl/min
gallons per minute, gal/min (gpm)	0.00223	cubic feet per second, ft^3/sec
gallons per minute, gal/min (gpm)	0.003785	cubic meters per minute, m^3/min
gallons per minute, gal/min (gpm)	0.0631	liters per second, l/sec

(*Continued*)

Common Conversion Factors (*Continued*)

Multiply	by	to Get
grams per cubic centimeter, g/cm^3	62.43	pounds per cubic foot, lb/ft^3
grams per cubic centimeter, g/cm^3	0.03613	pounds per cubic inch, lb/in^3
grams per liter, g/l	0.00834	pounds per gallon, lb/gal
hogsheads (U.S.)	8.422	cubic feet, ft^3
horsepower, hp	42.44	Btu per minute, Btu/min
horsepower, hp	746	joules per second, j/sec
horsepower, hp	746	watts, W
inches, in	2.54	centimeters, cm
inches, in	0.0833	feet, ft
inches, in	0.0254	meters, m
inches of mercury, in	0.0333	atmospheres, atm
inches of mercury, in	1.133	feet of water (head)
inches of mercury, in	0.03453	kilograms per sq centimeter, kg/cm^2
inches of mercury, in	0.4911	pounds per square inch, lb/in^2
inches of water, (4°C)	0.002455	atmospheres, atm
inches of water, (4°C)	0.0361	pounds per square inch, lb/in^2
joules, j	0.2391	calories (gram), g-cal
kilograms, kg	2.205	pounds, lb
kilograms, kg	0.001102	tons (short)
kilograms per cubic meter, kg/m^3	0.001	grams per cubic centimeter, g/cm^3
kilograms per cubic meter, kg/m^3	0.06243	pounds per cubic foot, lb/ft^3
kilograms per square centimeter, kg/cm^2	28.96	inches of mercury, in Hg
kilograms per square centimeter, kg/cm^2	32.81	feet of water, ft (head)
kilograms per square centimeter, kg/cm^2	14.22	pounds per square inch, lb/in^2
kilometers, km	0.6214	miles, mi (statute)
kilometers, km	0.5396	miles, NM (nautical)

(*Continued*)

Common Conversion Factors (*Continued*)

Multiply	by	to Get
kilometers per hour	27.78	centimeters per second, cm/sec
kilometers per hour	54.68	feet per minute, ft/min
kilometers per hour	0.6214	miles per hour, mi/hr
kilopascals, kPa	0.145	pounds per square inch, lb/in^2
kilowatts, kW	56.92	Btu per minute, Btu/min
kilowatts, kW	1.341	horsepower, hp
kilowatt-hours, kWh	860.5	kilogram-calories, kg-cal
liters, l	0.2642	gallons, gal
liters, l	0.00629	barrels (oilfield), bbl
liters, l	0.0353	cubic feet, ft^3
liters, l	0.001	cubic meters, m^3
liters, l	1.057	quarts (U.S.), qt
liters per minute, l/min	0.2642	gallons per minute, gal/min
liters per minute, l/min	0.00629	barrels per minute, bbl/min
meters, m	100	centimeters, cm
meters, m	3.281	feet, ft
meters, m	0.9144	yards, yd
meters per min, m/min	3.281	feet per minute, ft/min
meters per min, m/min	0.060	kilometers per hour, km/hr
meters per min, m/min	0.03728	miles per hour, mi/hr
miles (statute), mi	5280	feet, ft
miles (statute), mi	1609	meters, m
miles (statute), mi	1.609	kilometers, km
miles (statute), mi	0.8690	nautical miles, NM
miles per hour, mi/hr	1.466	feet per second, ft/sec
miles per hour, mi/hr	0.6214	kilometers per hour, km/hr
miles per hour, mi/hr	0.8690	knots, kn
miles per hour, mi/hr	26.82	meters per minute
nautical miles, NM	6076	feet, ft
nautical miles, NM	1852	meters, m
nautical miles, NM	1.151	statute miles, mi
newtons per square meter, N/m^2	1	pascals, Pa

(*Continued*)

Common Conversion Factors (*Continued*)

Multiply	by	to Get
ounces, oz	28.35	grams, g
ounces, oz	0.0625	pounds, lb
ounces, oz	0.9115	ounces (troy)
parts per million, ppm	0.0584	grains per gal (U.S.), grain/gal
parts per million, ppm	0.0001	weight percent, wt%
pounds, lb	453.6	grams, g
pounds, lb	0.4356	kilograms, kg
pounds, lb	16	ounces, oz
pounds per gallon, lb/gal	119.8	kilograms per cubic meter, kg/m^3
pounds per gallon, lb/gal	7.48	pounds per cubic foot, lb/ft^3
pounds per square inch, lb/in^2	2.307	feet of water (head)
pounds per square inch, lb/in^2	703.1	kilograms per square meter, kg/m^2
pounds per square inch, lb/in^2	6.897	kilopascals, kPa
pounds per square inch, lb/in^2	144	pounds per square foot, lb/ft^2
quarts (U.S.), qt	946.3	cubic centimeters, cm^3
quarts (U.S.), qt	0.9463	liters, l
quarts (U.S.), qt	0.0334	cubic feet, ft^3
quarts (U.S.), qt	57.75	cubic inches, in^3
radians	57.30	degrees
square centimeters, cm^2	0.001076	square feet, ft^2
square centimeters, cm^2	0.1550	square inches, in^2
square centimeters, cm^2	0.0001	square meters, m^2
square feet, ft^2	929	square centimeters, cm^2
square feet, ft^2	144	square inches, in^2
square feet, ft^2	0.0929	square meters, m^2
square inches, in^2	6.45	square centimeters, cm^2
square inches, in^2	0.00694	square feet, ft^2
square inches, in^2	0.000645	square meters, m^2
square kilometers, km^2	247.1	acres, ac

(*Continued*)

Common Conversion Factors (*Continued*)

Multiply	by	to Get
square kilometers, km^2	0.3861	square miles, mi^2
square meters, m^2	0.000247	acres, ac
square meters, m^2	10.76	square feet, ft^2
square meters, m^2	1.196	square yards, yd^2
square yards, yd^2	9	square feet, ft^2
square yards, yd^2	1296	square inches, in^2
square yards, yd^2	0.8361	square meters, m^2
tons (long)	1016	kilograms, kg
tons (long)	2240	pounds, lb
tons (long)	1.016	tons (metric)
tons (long)	1.120	tons (short)
tons (metric)	1000	kilograms, kg
tons (metric)	2204.6	pounds, lb
tons (metric)	0.9841	tons (long)
tons (metric)	1.1023	tons (short)
tons (short)	907.2	kilograms, kg
tons (short)	2000	pounds, lb
tons (short)	0.8929	tons (long)
tons (short)	0.907	tons (metric)
watts, W	3.415	Btu per hour, Btu/hr
watts, W	44.25	foot pounds per minute, ft-lb/min
watts, W	1	joules per second, j/sec
yards, yd	91.44	centimeters, cm
yards, yd	3	feet, ft
yards, yd	36	inches, in
yards, yd	0.000914	kilometers, km
yards, yd	0.9144	meters, m
yards, yd	0.000568	miles, mi

Conversion Factors: Metric Units to US Field Units

Units	Multiply by	Answer in
Meters [m)	× 3.2808	Feet
Centimeters (cm)	× 0.3937	Inches
Millimeters (mm)	× 0.03937	Inches
Metric Tons	× 2204.6	Pounds (Lbs)
Decanewtons (daN)	× 2.2481	Pounds (Lbs)
Kilograms	× 2.2046	Pounds
Kg/m	× 0.67196	Weight (Lbs/Ft)
Kg/m^3	× 0.3505	Pounds per Barrel
Liters	× 0.00629	Barrels
Cubic Meters	× 6.2898	Barrels
Liters	× 0.2642	Gallons
Cubic Meters	× 264.173	Gallons
Liters/Stroke	× 0.00629	Barrels/Stroke
Cubic Meters/Stroke	× 6.2898	Barrels/Stroke
Liters/Minute	× 0.2642	Gallons/Minute
Liters/Minute	× 0.00629	Barrels/Minute
Cubic Meters/Minute	× 6.2898	Barrels/Minute
Liters/Meter (l/m)	× 0.0019171	BBL/Ft. Capacity
Cubic Meters/Meter	× 1.917	BBL/Ft. Capacity
Liters/Meter (l/m)	× 0.0019171	BBL Displacement
Cubic Meters/Meter	× 1.9171	BBL Displacement
KPa/m	× 0.044207	Gradient PSI/Ft
Bar/m	× 4.4207	Gradient PSI/Ft
Kilograms/Liter (Kg/L)	× 8.3454	Mud Weight PPG
Kilograms/Cubic Meter	× 0.0083454	Mud Weight PPG
Specific Gravity (SG)	× 8.3454	Mud Weight PPG
Kg/m^3	× 6.24279	Mud Weight Lb/Ft^3)
Celsius Degrees	× 1.8 + 32	Fahrenheit Degrees
Pascals (Pa)	× 0.000145	PSI
Kilopascals (KPa)	× 0.14504	PSI
Bar	× 14.50377	PSI
Kg/Minute	× 8.475	BWPD @ 8.9 ppg

(Continued)

Conversion Factors: Metric Units to US Field Units (*Continued*)

Units	Multiply by	Answer in
Kg/Minute	× 10.105	BOPD @ 7.74 ppg
Kg/Minute	× 0.071	mmCFD @ 0.6 sp.gr.

Conversion Factors: US Field Units to Metric Units

Units	Multiply by	Answer in
Feet	× 0.3048	Meters (M)
Inches	× 2.54	Centimeters (cm)
Inches	× 25.4	Millimeters (mm)
Pounds (Lbs)	× 0.0004536	Metric Tons
Pounds (Lbs)	× 0.44482	Decanewtons (daN)
Pounds	× 0.4536	Kilograms
Weight (Lbs/ft)	× 1.4882	Kg/M
Pounds per Barrel	× 2.85307	Kg/M^3
Barrels	× 158.987	Liters
Barrels	× 0.15898	Cubic Meters
Gallons	× 3.7854	Liters
Gallons	× 0.0037854	Cubic Meters
Barrels/Stroke	× 158.987	Liters/Stroke
Barrels/Stroke	× 0.158987	Cubic Meters/Stroke
Gallons/Minute	× 3.7854	Liters/Minute
Barrels/Minute	× 158.987	Ljters/Minute
Banels/Minute	× 0.158987	Cubic Meters/Minute
bbl/ft. Capacity	× 521.612	Liters/Meter (L/M)
bbl/ft. Capacity	× 0.521612	Cubic Meters/Meter
Bbl Displacement	× 521.612	Liters/Meter (L/M)
Bbl Displacement	× 0.521612	Cubic Meters/Meter
Gradient psl/ft	× 22.6206	KPa/M
Gradient psi/ft	× 0.226206	Bar/M
Mud Weight PPG	× 0.119826	Kilograms/Liter (Kg/L)
Mud Weight PPG	× 119.826	Kilograms/Cubic Meter
Mud Weight PPG	×O.l 19826	Specific Gravity (SG)

(*Continued*)

Conversion Factors: US Field Units to Metric Units (*Continued*)

Units	Multiply by	Answer in
Mud Weight (Lb/Ft^3)	× 1.60185	Kg/M^3
Fahrenheit Degrees	× 0.56–17.8	Celsius Degrees
PSI	× 6894.8	Pascals (Pa)
PSI	× 6.8948	Kilopascals (KPa)
PSI	× 0.06895	Bar
BWPD @ 8.9 ppg	× 0.1 18	Kg/Min
BOPD ® 7.74 ppg	× 0.099	Kg/Min
mmCFD S' 0.6 sp. gr.	× 14.1	Kg/Min

Conversion Table – psi to kilopascal

1	psi = 6.894745 kPa	26	psi = 179.263365 kPa
2	psi = 13.78949 kPa	27	psi = 186.15811 kPa
3	psi = 20.684234 kPa	28	psi = 193.052855 kPa
4	psi = 27.578979 kPa	29	psi = 199.9476 kPa
5	psi = 34.473724 kPa	30	psi = 206.842345 kPa
6	psi = 41.368469 kPa	31	psi = 213.73709 kPa
7	psi = 48.263214 kPa	32	psi = 220.631834 kPa
8	psi = 55.157959 kPa	33	psi = 227.526579 kPa
9	psi = 62.052703 kPa	34	psi = 234.421324 kPa
10	psi = 68.947448 kPa	35	psi = 241.316069 kPa
11	psi = 75.842193 kPa	36	psi = 248.210814 kPa
12	psi = 82.736938 kPa	37	psi = 255.105559 kPa
13	psi = 89.631683 kPa	38	psi = 262.000303 kPa
14	psi = 96.526428 kPa	39	psi = 268.895048 kPa
15	psi = 103.421172 kPa	40	psi = 275.789793 kPa
16	psi = 110.315917 kPa	41	psi = 282.684538 kPa
17	psi = 117.210662 kPa	42	psi = 289.579283 kPa
18	psi = 124.105407 kPa	43	psi = 296.474027 kPa
19	psi = 131.000152 kPa	44	psi = 303.368772 kPa
20	psi = 137.894897 kPa	45	psi = 310.263517 kPa
21	psi = 144.789641 kPa	46	psi = 317.158262 kPa
22	psi = 151.684386 kPa	47	psi = 324.053007 kPa

(*Continued*)

Conversion Table – psi to kilopascal (*Continued*)

23	psi = 158.579131 kPa	48	psi = 330.947752 kPa
24	psi = 165.473876 kPa	49	psi = 337.842496 kPa
25	psi = 172.368621 kPa	50	psi = 344.737241 kPa

Conversion Table – psi to megapascal

psi	MPa	Psi	MPa	Psi	MPa	Psi	MPa
1	0.006894757	26	0.17926369	51	0.351632622	76	0.524001554
2	0.013789515	27	0.186158447	52	0.358527379	77	0.530896311
3	0.020684272	28	0.193053204	53	0.365422136	78	0.537791069
4	0.027579029	29	0.199947961	54	0.372316894	79	0.544685826
5	0.034473786	30	0.206842719	55	0.379211651	80	0.551580583
6	0.041368544	31	0.213737476	56	0.386106408	81	0.55847534
7	0.048263301	32	0.220632233	57	0.393001166	82	0.565370098
8	0.055158058	33	0.227526991	58	0.399895923	83	0.572264855
9	0.062052816	34	0.234421748	59	0.40679068	84	0.579159612
10	0.068947573	35	0.241316505	60	0.413685437	85	0.58605437
11	0.07584233	36	0.248211262	61	0.420580195	86	0.592949127
12	0.082737087	37	0.25510602	62	0.427474952	87	0.599843884
13	0.089631845	38	0.262000777	63	0.434369709	88	0.606738642
14	0.096526602	39	0.268895534	64	0.441264467	89	0.613633399
15	0.103421359	40	0.275790292	65	0.448159224	90	0.620528156
16	0.110316117	41	0.282685049	66	0.455053981	100	0.689475729
17	0.117210874	42	0.289579806	67	0.461948738	125	0.861844661
18	0.124105631	43	0.296474563	68	0.468843496	150	1.034213594
19	0.131000389	44	0.303369321	69	0.475738253	175	1.206582526
20	0.137895146	45	0.310264078	70	0.48263301	200	1.378951458
21	0.144789903	46	0.317158835	71	0.489527768	250	1.723689323
22	0.15168466	47	0.324053593	72	0.496422525	300	2.068427187
23	0.158579418	48	0.33094835	73	0.503317282	500	3.447378645
24	0.165474175	49	0.337843107	74	0.510212039	750	5.171067968
25	0.172368932	50	0.344737865	75	0.517106797	1000	6.89475729

Converting Milligrams per Liter to Parts per Million
To make the conversion, divide the value reported in mg/l by the specific gravity of the base fluid to convert to ppm.

$$(mg/1)/SG = ppm$$

A third common unit of concentration is weight percent (wt.%). Another way to think about weight percent is parts per hundred. By its definition, weight percent is a ratio of pounds of a substance per hundred pounds of total weight. It is a ratio of weight per weight and, as such, is similar to ppm. The easiest way to remember the conversion from weight percent to parts per million is that one tenth of a percent (0.1%) by weight is equal to 1,000 parts per million.

$$0.1 \text{ wt.}\% = 1{,}000 \text{ ppm}$$

A quick conversion between the three different units of concentration can be made if the specific gravity is known.

Drill Pipe Capacity and Displacement

Nominal Size inches	Weight lb/ft	OD inches	ID inches	Capacity bbl/ft	Capacity ft/bbl	Pipe Displacement bbl/100 ft
2-3/8	6.65	2.375	1.815	0.0032	311.43	0.228
2-7/8	10.40	2.875	2.151	0.0045	221.73	0.353
3-1/2	9.50	3.500	2.992	0.0087	114.60	0.320
3-1/2	13.30	3.500	2.764	0.0074	134.29	0.448
3-1/2	15.50	3.500	2.602	0.0066	151.53	0.532
4	11.85	4.000	3.476	0.0118	84.91	0.381
4	14.00	4.000	3.340	0.0109	91.96	0.471
4-1/2	13.75	4.500	3.958	0.0153	65.49	0.445
4-1/2	16.60	4.500	3.826	0.0143	70.08	0.545
4-1/2	20.00	4.500	3.640	0.0129	77.43	0.680
5	16.25	5.000	4.408	0.0189	52.80	0.541
5	19.50	5.000	4.276	0.0178	56.11	0.652
5	25.60	5.000	4.000	0.0156	64.12	0.874

(*Continued*)

Drill Pipe Capacity and Displacement (*Continued*)

Nominal Size inches	Weight lb/ft	OD inches	ID inches	Capacity bbl/ft	Capacity ft/bbl	Pipe Displacement bbl/100 ft
5-1/2	21.90	5.500	4.778	0.0223	44.94	0.721
5-1/2	24.70	5.500	4.670	0.0213	47.04	0.820

Geological Time Scale

There are various geologic eras and periods that correlate with oil-producing formations throughout the world. Only the Cretaceous period of the Mesozoic Era, and the Paleocene, Eocene, Oligocene, Miocene, Pliocene, and Pleistocene are, or have been, significantly represented in oil fields of the United States.

Table The Geological Time Scale.

Era	Period	Epoch	Duration (years x 106)*	Years ago (years x 106)*
Cenozoic	Quaternary	Holocene	<10,000 years ago	
		Pleistocene	2	.01
	Tertiary	Pliocene	11	2
		Miocene	12	13
		Oligocene	11	25
		Eocene	22	36
		Paleocene	71	58
Mesozoic	Cretaceous		71	65
	Jurassic		54	136
	Triassic		35	190
Paleozoic	Permian		55	225
	Carboniferous		65	280
	Devonian		60	345
	Silurian		20	405
	Ordovician		75	425
	Cambrian		100	500
Precambrian			3,380	600

*Approximations.

Hole Capacity

Hole diameter inches	Capacity bbl/ft	Capacity ft/bbl
3	0.0087	114.387
3-1/8	0.0095	105.419
3-1/4	0.0103	97.466
3-3/8	0.0111	90.380
3-1/2	0.0119	84.040
3-5/8	0.0128	78.344
3-3/4	0.0137	73.208
3-7/8	0.0146	68.561
4	0.0155	64.343
4-1/8	0.0165	60.502
4-1/4	0.0175	56.996
4-3/8	0.0186	53.785
4-1/2	0.0197	50.839
4-5/8	0.0208	48.128
4-3/4	0.0219	45.628
4-7/8	0.0231	43.318
5	0.0243	41.179
5-1/8	0.0255	39.195
5-1/4	0.0268	37.351
5-3/8	0.0281	35.634
5-1/2	0.0294	34.033
5-5/8	0.0307	32.537
5-3/4	0.0321	31.138
5-7/8	0.0335	29.827
6	0.0350	28.597
6-1/8	0.0364	27.442
6-1/4	0.0379	26.355
6-3/8	0.0395	25.331
6-1/2	0.0410	24.367
6-5/8	0.0426	23.456
6-3/4	0.0443	22.595

(*Continued*)

Hole Capacity (*Continued*)

Hole diameter inches	Capacity bbl/ft	Capacity ft/bbl
6-7/8	0.0459	21.781
7	0.0476	21.010
7-1/8	0.0493	20.279
7-1/4	0.0511	19.586
7-3/8	0.0528	18.928
7-1/2	0.0546	18.302
7-5/8	0.0565	17.707
7-3/4	0.0583	17.140
7-7/8	0.0602	16.600
8	0.0622	16.086
8-1/8	0.0641	15.595
8-1/4	0.0661	15.126
8-3/8	0.0681	14.677
8-1/2	0.0702	14.249
8-5/8	0.0723	13.839
8-3/4	0.0744	13.446
8-7/8	0.0765	13.070
9	0.0787	12.710
9-1/8	0.0809	12.364
9-1/4	0.0831	12.032
9-3/8	0.0854	11.713
9-1/2	0.0877	11.407
9-5/8	0.0900	11.113
9-3/4	0.0923	10.830
9-7/8	0.0947	10.557
10	0.0971	10.295
10-1/8	0.0996	10.042
10-1/4	0.1021	9.799
10-3/8	0.1046	9.564
10-1/2	0.1071	9.338
10-5/8	0.1097	9.119

(*Continued*)

Hole Capacity (*Continued*)

Hole diameter inches	Capacity bbl/ft	Capacity ft/bbl
10-3/4	0.1123	8.908
10-7/8	0.1149	8.705
11	0.1175	8.508
11-1/8	0.1202	8.318
11-1/4	0.1229	8.134
11-3/8	0.1257	7.956
11-1/2	0.1285	7.784
11-5/8	0.1313	7.618
11-3/4	0.1341	7.457
11-7/8	0.1370	7.301
12	0.1399	7.149
12-1/8	0.1428	7.003
12-1/4	0.1458	6.860
12-3/8	0.1488	6.722
12-1/2	0.1518	6.589
12-5/8	0.1548	6.459
12-3/4	0.1579	6.333
12-7/8	0.1610	6.210
13	0.1642	6.092
13-1/8	0.1673	5.976
13-1/4	0.1705	5.864
13-3/8	0.1738	5.755
13-1/2	0.1770	5.649
13-5/8	0.1803	5.546
13-3/4	0.1836	5.445
13-7/8	0.1870	5.348
14	0.1904	5.252
14-1/8	0.1938	5.160
14-1/4	0.1972	5.070
14-3/8	0.2007	4.982
14-1/2	0.2042	4.896

(*Continued*)

Hole Capacity (*Continued*)

Hole diameter inches	Capacity bbl/ft	Capacity ft/bbl
14-5/8	0.2078	4.813
14-3/4	0.2113	4.732
14-7/8	0.2149	4.653
15	0.2186	4.575
15-1/8	0.2222	4.500
15-1/4	0.2259	4.427
15-3/8	0.2296	4.355
15-1/2	0.2334	4.285
15-5/8	0.2371	4.217
15-3/4	0.2410	4.150
15-7/8	0.2448	4.085
16	0.2487	4.021
16-1/8	0.2526	3.959
16-1/4	0.2565	3.899
16-3/8	0.2605	3.839
16-1/2	0.2645	3.781
16-5/8	0.2685	3.725
16-3/4	0.2725	3.669
16-7/8	0.2766	3.615
17	0.2807	3.562
17-1/4	0.2890	3.460
17-1/2	0.2975	3.362
17-3/4	0.3060	3.268
18	0.3147	3.177
18-1/4	0.3235	3.091
18-1/2	0.3324	3.008
18-3/4	0.3415	2.928
19	0.3507	2.852
19-1/4	0.3599	2.778
19-1/2	0.3694	2.707
19-3/4	0.3789	2.639
20	0.3885	2.574

(*Continued*)

Hole Capacity (*Continued*)

Hole diameter inches	Capacity bbl/ft	Capacity ft/bbl
20-1/4	0.3983	2.511
20-1/2	0.4082	2.450
20-3/4	0.4182	2.391
21	0.4284	2.334
21-1/4	0.4386	2.280
21-1/2	0.4490	2.227
21-3/4	0.4595	2.176
22	0.4701	2.127
22-1/4	0.4809	2.080
22-1/2	0.4918	2.034
22-3/4	0.5027	1.989
23	0.5138	1.946
23-1/4	0.5251	1.904
23-1/2	0.5364	1.864
23-3/4	0.5479	1.825
24	0.5595	1.787
24-1/4	0.5712	1.751
24-1/2	0.5831	1.715
24-3/4	0.5950	1.681
25	0.6071	1.647
26	0.6566	1.523
27	0.7081	1.412
28	0.7615	1.313
29	0.8169	1.224
30	0.8742	1.144
31	0.9335	1.071
32	0.9947	1.005

Physical Constants

Name	Value	Unit
Absolute zero	-273.15	°C
Acceleration of free fall	9.80665	m s-2
Atomic mass unit	$1.660538782(83) \times 10^{-27}$	Kg
Avogadro constant	$6.02214179(30) \times 10^{23}$	mol-1
Base of natural logarithms	2.718 281828459	
Bohr magneton	$927.400915(23) \times 10^{-26}$	J T-1
Bohr radius	$0.52917720859(36) \times 10^{-10}$	m
Boltzmann constant	$1.3806504(24) \times 10^{-23}$	J K-1
Characteristic impedance of vacuum	376.730313461	Ω
Classical electron radius	$2.8179402894(58) \times 10^{-15}$	m
Dirac constant	$1.054571628(53) \times 10^{-34}$	J s
Electron mass	$9.10938215(45) \times 10^{-31}$	kg
Electron-proton mass ratio	$5.4461702177(24) \times 10^{-4}$	
Electron volt	$1.602176487(40) \times 10^{-19}$	J
Elementary charge	$1.602176487(40) \times 10^{-19}$	C
Euler's number	2.718281828459...	
Faraday constant	96485.3399(24)	C mol-1
Feigenbaum constant	4.669201609102990...	
Feigenbaum reduction parameter	2.502907875095892...	
Fine-structure constant	7.2973525376 (50) × 10-3	
First radiation constant	$3.74177118(19) \times 10^{-16}$	W m2
Hartree energy	4.359743 94(22) ×10-18	J
Josephson constant	$483597.891(12) \times 10^{9}$	Hz V-1
Loschmidt constant	2.686 7774(47) × 1025	m-3
Magnetic flux quantum	$2.067833667(52) \times 10^{-15}$	Wb
Molar gas constant	8.314472(15)	J mol-1K-1
Molar Planck constant	$3.9903126821(57) \times 10^{-10}$	J s mol-1
Molar volume (Ideal gas, T = 273.15 K, p = 101.325 kPa)	$22.413996(39) \times 10^{-3}$	m3mol-1
Neutron mass	$1.674927211(84) \times 10^{-27}$	kg
Neutron-proton mass ratio	1.00137841918(46)	

(Continued)

Physical Constants (*Continued*)

Name	Value	Unit
Newtonian constant of gravitation	$6.67428(67) \times 10^{-11}$	m3 kg-1s-2
Nuclear magneton	$5.05078324(13) \times 10^{-27}$	J T-1
Permeability of vacuum	12.566370614 × 10-7	N A-2
Permittivity of vacuum	8.854187817 × 10-12	F m-1
Pi	3.141592653589793238	
Planck constant	$6.62606896(33) \times 10^{-34}$	J s
Proton mass	$1.672621637(83) \times 10^{-27}$	kg
Reduced Planck constant	$1.054571628(53) \times 10^{-34}$	J s
Rydberg constant	10973731.568527(73)	m-1
Second radiation constant	$1.4387752(25) \times 10^{-2}$	m K
Solar constant	1366	W m-2
Speed of light in vacuum	299792458	m s-1
Speed of sound in air	331.5 + 0.6 * T/°C	m s-1
Standard pressure	101325	Pa
Stefan-Boltzmann constant	$5.670400(40) \times 10^{-8}$	W m-2K-4

Production Engineering Units

Variable	Oilfield units	SI	Conversion (multiply oilfield unit)
Area	acre	nr	4,04 * icy^{1}
Compressibility	psL^{1}	Pa^{1}	] 45 × 1 d^{4}
Length	ft	m	3,05 X 10-^{L}
Permeability	md	nr	9,9 × ia^{lfi}
Pressure	psi	Pa	6,9 × 10^{s}
Rate (oil)	stb/d	mVs	1.84 x
Rate (gas)	Mscf/d	m^3/s	3,28 × Id^{4}
Viscosity	cp	Pa-s	i × m^{3}

SI Units

SI base units

Base quantity		SI base unit	
Name of base quantity	**Symbol**	**Name of SI base unit**	**Symbol**
Length	l, x, r,etc.	metre	m
Mass	m	kilogram	kg
time, duration	t	second	s
electric current	I, i	ampere	A
thermodynamic temperature	T	kelvin	K
amount of substance	n	mole	mol
luminous intensity	I_v	candela	cd

Examples of coherent derived units in the SI expressed in terms of base units

Derived quantity		SI coherent derived unit	
Name	**Symbol**	**Name**	**Symbol**
Area	A	square metre	m^2
volume	V	cubic metre	m^3
speed, velocity	v	metre per second	$m\ s^{-1}$
acceleration	a	metre per second squared	$m\ s^{-2}$
wavenumber	σ	reciprocal metre	m^{-1}
density, mass density	ρ	kilogram per cubic metre	$kg\ m^{-3}$
surface density	ρ_A	kilogram per square metre	$kg\ m^{-2}$
specific volume	v	cubic metre per kilogram	m^3kg^{-1}
current density	j	ampere per square metre	$A\ m^{-2}$
magnetic field strength	H	ampere per metre	$A\ m^{-1}$
amount concentration, concentration	c	mole per cubic metre	$mol\ m^{-3}$
mass concentration	ρ, γ	kilogram per cubic metre	$kg\ m^{-3}$
luminance	L_v	candela per square metre	$cd\ m^{-2}$
refractive index	n	(the number) one	1
relative permeability	μ_r	(the number) one	1

Coherent derived units in the SI with special names and symbols

	SI coherent derived unit			
Derived quantity	**Name**	**Symbol**	**Expressed in terms of other SI units**	**Expressed in terms of SI base units**
plane angle	radian	rad	1	$m\ m^{-1}$
solid angle	steradian	sr	1	$m^2\ m^{-2}$
frequency	hertz	Hz		s^{-1}
Force	newton	N		$m\ kg\ s^{-2}$
pressure, stress	pascal	Pa	N/m^2	$m^{-1}\ kg\ s^{-2}$
energy, work, amount of heat	joule	J	N m	$m^2\ kg\ s^{-2}$
power, radiant flux	watt	W	J/s	$m^2\ kg\ s^{-3}$
electric charge, amount of electricity	coulomb	C		s A
electric potential difference, electromotive force	volt	V	W/A	$m^2\ kg\ s^{-3}A^{-1}$
capacitance	farad	F	C/V	$m^{-2}\ kg^{-1}\ s^4A^2$
electric resistance	ohm	Ω	V/A	$m^2\ kg\ s^{-3}A^{-2}$
electric conductance	siemens	S	A/V	$m^{-2}\ kg^{-1}\ s^3A^2$
magnetic flux	weber	Wb	V s	$m^2\ kg\ s^{-2}A^{-1}$
magnetic flux density	tesla	T	Wb/m^2	$kg\ s^{-2}\ A^{-1}$
inductance	henry	H	Wb/A	$m^2\ kg\ s^{-2}A^{-2}$
Celsius temperature	degree C	°C		K
luminous flux	lumen	lm	cd sr	cd
luminance	lux	lx	lm/m^2	m^{-2} cd
activity referred to a radionuclide	becquerel	Bq		s^{-1}

Examples of SI coherent derived units whose names and symbols include SI coherent derived units with special names and symbols

		SI coherent derived unit	
Derived quantity	**Name**	**Symbol**	**Expressed in terms of SI base units**
dynamic viscosity	pascal second	Pa s	m^{-1} kg s^{-1}
moment of force	newton metre	N m	m^2 kg s^{-2}
surface tension	newton per metre	N/m	kg s^{-2}
angular velocity	radian per second	rad/s	m m^{-1} s^{-1} = s^{-1}
angular acceleration	radian per second squared	rad/s^2	m m^{-1} s^{-2} = s^{-2}
heat flux density, irradiance	watt per square metre	W/m^2	kg s^{-3}
heat capacity, entropy	joule per kelvin	J/K	m^2 kg s^{-2} K^{-1}
specific heat capacity, specific entropy	joule per kilogram kelvin	J/(kg K)	m^2 s^{-2} K^{-1}
specific energy	joule per kilogram	J/kg	m^2 s^{-2}
thermal conductivity	watt per metre kelvin	W/(m K)	m kg s^{-3} K^{-1}
energy density	joule per cubic metre	J/m^3	m^{-1} kg s^{-2}
electric field strength	volt per metre	V/m	m kg s^{-3} A^{-1}
electric charge density	coulomb per cubic metre	C/m^3	m^{-3} s A

surface charge density	coulomb per square metre	C/m^2	m^{-2} s A
electric flux density, electric displacement	coulomb per square metre	C/m^2	m^{-2} s A
permittivity	farad per metre	F/m	$m^{-3}\ kg^{-1}\ s^4\ A^2$
permeability	henry per metre	H/m	$m\ kg\ s^{-2}\ A^{-2}$
molar energy	joule per mole	J/mol	$m^2\ kg\ s^{-2}\ mol^{-1}$
molar entropy, molar heat capacity	joule per mole kelvin	J/(mol K)	$m^2\ kg\ s^{-2}\ K^{-1}\ mol^{-1}$
exposure (x- and γ-rays)	coulomb per kilogram	C/kg	kg^{-1} s A
absorbed dose rate	gray per second	Gy/s	$m^2\ s^{-3}$
radiant intensity	watt per steradian	W/sr	$m^4\ m^{-2}\ kg\ s^{-3} = m^2\ kg\ s^{-3}$
radiance	watt per square metre steradian	$W/(m^2sr)$	$m^2\ m^{-2}\ kg\ s^{-3} = kg\ s^{-3}$
catalytic activity concentration	katal per cubic metre	kat/m^3	$m^{-3}\ s^{-1}$ mol

Non-SI units accepted for use with the International System of Units

Quantity	Name of unit	Symbol for unit	Value in SI units
time, duration	**minute**	**min**	**1 min = 60 s**
	hour	h	1 h = 60 min = 3 600 s
	day	d	1 d = 24 h = 86 400 s
plane angle	degree	°	1° = (π/180) rad
	minute	‘	1’ = (1/60)° = (π/10 800) rad
	second	“	1” = (1/60)’ = (π/648 000) rad
area	hectare	ha	1 ha = 1hm^2 = 10^4 m^2
volume	litre	L, l	1 L = 1 dm^3 = 10^{-3} m^3
mass	tonne	t	1 t = 10^3 kg

Non-SI units whose values in SI units must be obtained experimentally

Quantity	Name of unit	Symbol for unit	Value in SI units
Units accepted for use with the SI			
energy	electronvolt	eV	1 eV = 1.602 176 53(14)×10^{-19} J
mass	dalton	Da	1 Da = 1.660 538 86(28)×10^{-27} kg
	unified atomic mass unit	u	1 u = 1 Da
length	astronomical unit	ua	1 ua = 1.495 978 706 91(6)×10^{11} m
Natural units (n.u.)			
speed, velocity	natural unit of speed (speed of light in vacuum)	c_o	299 792 458 m s^{-1}
action	natural unit of action (reduced Planck constant)	$\hbar$	1.054 571 68(18)×10-34 Js
mass	natural unit of mass (electron mass)	m_e	9.109 382 6(16)×10-31 kg
time, duration	natural unit of time	$\hbar/(m_e c_o^2)$	1.288 088 667 7(86)×10-21s

(Continued)

Non-SI units whose values in SI units must be obtained experimentally (*Continued*)

Quantity	Name of unit	Symbol for unit	Value in SI units
Atomic units (a.u.)			
charge	atomic unit of charge, (elementary charge)	e	1.602 176 53(14)×10-19 C
mass	atomic unit of mass, (electron mass)	m_e	9.109 382 6(16)×10-31 kg
action	atomic unit of action, (reduced Planck constant)	$\hbar$	1.054 571 68(18)×10-34 Js
length	atomic unit of length, bohr (Bohr radius)	a_o	0.529 177 210 8(18)×10-10m
energy	atomic unit of energy, hartree (Hartree energy)	E_h	4.359 744 17(75)×10-18 J
time, duration	atomic unit of tim	$\hbar/E_h$	2.418 884 326 505(16)×10-17 s

Other non-SI units

Quantity	Name of unit	Symbol for unit	Value in SI units
pressure	bar	bar	1 bar = 0.1 MPa = 10^5Pa
	millimetre of mercury	mmHg	1 mmHg ≈ 133.322 Pa
length	angström	Å	1 Å = 0.1 nm = 10^{-10} m
distance	nautical mile	M	1 M = 1852 m
area	barn	b	1 b = 100 fm^2 = 10^{-28} m^2
speed, velocity	knot	kn	1 kn = (1852/3600) m s^{-1}
logarithmic ratio quantities	neper	Np	
	bel	B	
	decibel	dB	

Non-SI units associated with the CGS system of units

Quantity	Name of unit	Symbol for unit	Value in SI units
energy	erg	erg	1 erg = 10^{-7} J
force	dyne	dyn	1 dyn = 10^{-5} N
dynamic viscosity	poise	P	1 P = 1 dyn s cm^{-2} = 0.1 Pa s
kinematic viscosity	stokes	St	1 St = 1 cm^2 s^{-1} = 10^{-4} m^2 s^{-1}
luminance	stilb	sb	1 sb = 1 cd cm^{-2} = 10^4 cd m^{-2}
illuminance	phot	ph	1 ph = 1 cd sr cm^{-2} = 10^4 lx
acceleration	gal	Gal	1 Gal = 1 cm s^{-2} = 10^{-2} m s^{-2}
magnetic flux	maxwell	Mx	1 Mx = 1 G cm^2 = 10^{-8} Wb
magnetic flux density	gauss	G	1 G = 1 Mx cm^{-2} = 10^{-4} T
magnetic field	œrsted	Oe	1 Oe ≙ ($10^3/4\pi$) A m^{-1}

SI prefixes

Factor	Name	Symbol	Factor	Name	Symbol
10^1	deca	da	10^{-1}	deci	d
10^2	hecto	h	10^{-2}	centi	c
10^3	kilo	k	10^{-3}	milli	m
10^6	mega	M	10^{-6}	micro	µ
10^9	giga	G	10^{-9}	nano	n
10^{12}	tera	T	10^{-12}	pico	p
10^{15}	peta	P	10^{-15}	femto	f
10^{18}	exa	E	10^{-18}	atto	a
10^{21}	zetta	Z	10^{-21}	zepto	z
10^{24}	yotta	Y	10^{-24}	yocto	y

Names and symbols for decimal multiples and submultiples of the unit of mass are formed by attaching prefix names to the unit name 'gram', and prefix symbols to the unit symbol 'g'.

These SI prefixes refer strictly to powers of 10. They should not be used to indicate powers of 2 (for example, one kilobit represents 1000 bits and not 1024 bits). The names and symbols for the prefixes corresponding to 2^{10}, 2^{20}, 2^{30}, 2^{40}, 2^{50}, and 2^{60} are, respectively: kibi, Ki; mebi, Mi; gibi, Gi; tebi, Ti; pebi, Pi; and exbi, Ei. Thus, for example, one kibibyte would be written: 1 KiB = 2^{10} B = 1024 B, where B denotes a byte. Although these prefixes are not part of the SI, they should be used in the field of information technology to avoid the incorrect usage of the SI prefixes.

Sources:

SPE. 1982. The SI Metric System of Units and SPE Metric Standard. Society of Petroleum Engineers, Richardson, Texas.

The International System of Units (SI). Bureau International des Poids et Mesures, Paris, France. 30 Nov 2010.

The International System of Units from NIST. Oct 2000. National Institute of Standards and Technology. 30 Nov 2010. <http://physics.nist.gov/cuu/Units/>.

Decimal Multipliers for SI Prefixes

Prefix	Origin	Symbol	Multiplying factor
yotla	Greek or Latin *ocio*, "eight"	Y	10^{24}
zetta	Latin *sepictn*, "seven"	Z	10^{2J}
cx.a	Greek *hex*, "six"	E	10^{18}
pet a	Greek *peme*. "five" '	P	10^{15}
tcra	Greek *reras*. "monster"	T	10^{12}
gig*	Greek *gigas*, "giant"	G	10^{9}
mega	Greek *megas*, "large"	M	10^{6}
kilo	Greek *chilioi*. "thousand"	k	10^{3}
hccto	Greek *hekeson*, "hundred"	h	10^{2}
delta	Greek *deka*, "ten"	da	10^{1}
dcci	Latin *decimui*, "fen th"	d	10^{-1}
centi	Latin *cemum*, "hundred"	c	10^{-2}
mi lb	Latin *mille*, "thousand"	m	10^{-3}
micro	Latin *micro-* (Greek mUcros), "small"	µ	10^{-6}
nano	Latin *nanus* (Greek flints j), " dAarf"	n	10^{-9}
pico	Spanish *pico*, "a bit," Italian *piccolo*, "small"	P	10^{-12}
femto	Danish-Norwc gi an *femten*," fifteen"	f	10^{-15}
alio	Danish-Norwegian *atten*, " eighteen"	a	10^{-18}

(*Continued*)

Decimal Multipliers for SI Prefixes (*Continued*)

Prefix	Origin	Symbol	Multiplying factor
zepto	Latin *septem*, "seven"	z	10^{-21}
yoelo	Greek or Latin ocio, "eight"	y	10^{-24}

Common Unit Conversions

Gas Constant 0.082 lilcr-aLm/mole K 62.26 Jiler-mm Hg/ mole K 8.314 Joule/g-mole K 1.314 atm-ft^3/lb-mole K 1.987 cal/g-mole K 1.987 Btu/lb-mole °R 0.73 atm-ft^3/lb-mole °R 10.73 psi-fl^3/lb-mole °R 1545 ft-lb$_f$/lb-mole °R	Volume 1 ft^3 = 28.316 liter = 7.481 gal 1 in.^{3} = 16.39 cc = 5.787×10^{-4} ft^3 1 gal = 3.785 liter = 8.34 lb H_2O 1 m^3 = 35.32 ft^3 = 264.2 gal	Density 1 g/cm^3 = 1000kg/m^3 = 62.428 Ib/ft^3 = 8.345 lb/gal = 003613 lb/in.^{3}
Length 1 mile = 1609 m = 5280 ft 1 fl = 30.48 cm = 12 in. 1 in. = 2.54 cm 1 im = 3.2808 ft = 39.37 in. 1 nm= 10^{-9} m = 10 A	Viscosity 1 poise = 6.7197×10^{-2} lb$_m$/ft-sec = 2.0886×10^{-3} lb$_f$-sec/ft^2 = 2.4191×10^2 lb$_m$/ft-hr = 1 g/cm-scc	Conversion Factor 1 cal/g-mole = l.8Btu/ lb-mole 1 amu = 1.66063×10^{-24}g 1 eV= 1.6022×10^{-24}erg 1 radian = 57.3° 1 cm/sec= 1.9685 ft/min 1 rpm= 0.10472 ratlian/ sec
Pressure 1 aim = 101325 N/m^2 = 14.696 psi = 760 mmHg = 29.921 in Hg (32 °F) = 33.91 ftH_2O (39.1 °F) = 2116.2 lb$_f$/ft^2 = 1.0133 bar = 1033.3 g$_f$/cm^2	Constant h = 6.6262×10^{-27}erg-sec k = 1.38062×10^{-16} erg/K N_0= 6.022169×10^{23} C= 2.997925×10^{10} cm/ sec F= 96487 coul/eq e = 1.60219×10^{-19} coul g = 980.665 cm/sec^2 =32.174 ft/sec^2	Mass 1 kg = 2.2046 lb 1 lb = 453.59 g 1 ton = 2000 lb = 907.2 kg 1 B ton = 2240 lb = 1016 kg 1 tonne = 2205 lb = 1000 kg 1 slug = 32.2 lb = 14.6 kg

(*Continued*)

Common Unit Conversions (*Continued*)

Area 1 m^2= 10.76 ft^2 = 1550 in.^{2} 1 ft^2 = 929.0 cm^2	Power 1 HP= 550 ft-lb$_f$/sec = 745.48 watt 1 Btu/hr = 0.293 watt	Force 1 N = 1 kg-m/sec^2 = 10^2 dyne = 0.22481 lb$_f$ = 7.233 lb$_m$-ft/sec^2
Transfer Coefficient 1 Btu/hr-ft^2 °F = 5.6784 Joule/sec-m^2 = 4.8825 Kcal/hr-m^2 K = 0.45362 Kcal/hr-ft^2 K = 1.3564 × 10^{-4} cal/sec-cm^2 K 1 lb/hr-ft^2 = 1.3562 × 10^3 kg/sec-m^2 = 4.8823 kg/hr-m^2 = 0.45358 kg/hr-ft^2 1 cal/g °C = 1 Btu/Ib$_m$ °F = 1 Pcu/lb$_m$ °C 1 Btu/hr-ft °F = 1.731 W/mK = 1.4882 kcal/hr-m K	Energy & Work 1 cal = 4.184 Joule 1 Btu = 1055.1 Joule = 252.16 cal 1 HP-hr = 2684500 Joule = 641620 cal = 2544.5 Btu 1 KW-hr = 3.6 × 10^6 Joule = 860565 cal = 3412.75 Blu 1 l-atm = 24.218 cal 1 fl-lb$_f$= 0.3241 cal 1 Pcu = 453.59 cal 1 kg-m = 2.3438 cal	Stress 1 MPa = 145 psi 1 MPa = 0.102 kg/mm^2 1 Pa= 10 dynes/cm^2 1 kg/mm^2= 9806 MPa 1 psi = 6.90 × 10^{-3} MPa 1 kg/mm^2= 9.806 MPa 1 dyne/cm^2 = 0.10 Pa 1 psi = 7.03 × 10^4 kg/mm^2 1 psi in.$^{1/2}$ = 1.099 × 10^{-3} MPa m$^{1/2}$ 1 MPa m$^{1/2}$ = 910 psi in.$^{1/2}$

Water – Boiling Point Variation with Pressure

psia	Boiling point, °F	psia	Boiling point, °F	psia	Boiling point, °F
0.5	79.6	44	273.1	150	358.5
1	101.7	46	275.8	175	371.8
2	126	48	278.5	200	381.9
3	141.4	50	281	225	391.9
4	125.9	52	283.5	250	401
5	162.2	54	285.9	275	409.5
6	170	56	288.3	300	417.4
7	176.8	58	290.5	325	424.8
8	182.8	60	292.7	350	431.8
9	188.3	62	294.9	375	438.4
10	193.2	64	297	400	444.7
11	197.7	66	299	425	450.7
12	201.9	68	301	450	456.4

(*Continued*)

Water – Boiling Point Variation with Pressure (*Continued*)

psia	Boiling point, °F	psia	Boiling point, °F	psia	Boiling point, °F
13	205.9	70	303	475	461.9
14	209.6	72	304.9	500	467.1
14.69	212	74	306.7	525	472.2
15	213	76	308.5	550	477.1
16	216.3	78	310.3	575	481.8
17	219.4	80	312.1	600	486.3
18	222.4	82	313.8	625	490.7
19	225.2	84	315.5	650	495
20	228	86	317.1	675	499.2
22	233	88	318.7	700	503.2
24	237.8	90	320.3	725	507.2
26	242.3	92	321.9	750	511
28	246.4	94	323.4	775	514.7
30	250.3	96	324.9	800	518.4
32	254.1	98	326.4	825	521.9
34	257.6	100	327.9	850	525.4
36	261	105	331.4	875	528.8
38	264.2	110	334.8	900	532.1
40	267.3	115	338.1	950	538.6
42	270.2	120	341.3	1000	544.8

Water – Density and Viscosity in Relation to Temperature

T, °C	Density, kgm/cubic meter	Viscosity, mPa
0	1000	1.788
10	1000	1.307
20	998	1.003
30	996	0.799
40	992	0.657
50	988	0.548
60	983	0.467
70	978	0.405

(*Continued*)

Water – Density and Viscosity in Relation to Temperature (*Continued*)

T, °C	Density, kgm/cubic meter	Viscosity, mPa
80	972	0.355
90	965	0.316
100	958	0.283

Water Temperature and Vapor Pressure

Temperature		Vapor Pressure	
F	C	psi	ll
40	4.4	0.12	0.28
50	10	0.18	0.41
60	15.6	0.26	0.59
70	21.1	0.36	0.82
SO	26.7	0.51	1.17
90	32.2	0.70	1.61
100	37.8	0.95	2.19
110	43.3	1.28	2.94
120	48.9	1.69	3.91
130	54.4	2.22	5.15
140	60	2.89	6.68
150	65.6	3.72	8.56
160	71.1	4.74	10.95
170	76.7	5.99	13.84
I8O	82.2	7.51	17.35
190	87.8	9.34	21.55
200	93.3	11.50	26.65
212	100	14.70	33.96

Weights and Measures – General

Unit of Measure	Equivalent
Barrel	31.5 gallons
Bushel	2150.4 cubic inches
Bushel (dry)	4 pecks
Centimeter	0.39 inches

(*Continued*)

Weights and Measures – General (*Continued*)

Unit of Measure	Equivalent
Cord	128 cubic feet
Cord	4 cubic meters
Cubic centimeter	0.061 cubic inches
Cubic foot	283.17 cubic centimeters
Cubic foot	1728 cubic inches
Cubic inch	16.39 cubic centimeters
Cubic meter	1.31 cubic yards
Cubic yard^3	0.77 cubic meters
Foot	30.48 centimeters
Gallon (us)	231 cubic inches
Grain	0.002 ounces
Gram	0.04 ounces
Hectare	2.47 acres
Hogshead	2 barrels
Inch	2.54 centimeters
Kilogram	2.21 pounds
Meter	1.09 yards
Mile	1.61 kilometers
Ounce	28.35 grams
Peck	8 quarts
Pound	0.45 kilograms
Pound (troy)	12 ounces
Square centimeter	0.16 square inches
Square foot	0.09 square meters
Square inch	6.45 square centimeters
Square meter	1.2 square yards
Square mile	2.59 square kilometers
Square yard	square 0.84 meters
Ton (long)	2240 pounds
Ton (metric)	2200 pounds
Ton (shipping)	40 feet^3
Ton (short)	2000 pounds

(*Continued*)

Weights and Measures – General (*Continued*)

Unit of Measure	Equivalent
Township	36 square miles (93 square kilometers)
Yard	0.91 meters

Glossary

Abnormal pressure: A formation pore pressure that is higher than that resulting from a water gradient.

Absolute temperature: Temperature related to absolute zero, the temperature at which all molecular activity ceases; calculated by adding 460°F to the temperature in Fahrenheit to obtain the absolute temperature in degrees Rankine or by adding 273°C to the temperature in degrees Celsius to obtain the absolute temperature in degrees Kelvin.

Absorption: The penetration or apparent disappearance of molecules or ions of one or more substances into the interior of a solid or liquid.

Accumulator: A hydraulic accumulator is a pressure storage reservoir in which a non-compressible hydraulic fluid is held under pressure that is applied by an external source; an accumulator enables a hydraulic system to cope with extremes of demand using a less powerful pump, to respond more quickly to a temporary demand, and to smooth out pulsations; compressed gas accumulators, also called hydro-pneumatic accumulators, are the most common type of accumulators.

Acid: Any chemical compound containing hydrogen capable of being replaced by elements or radicals to form salts; a compound, which, on dissociation in solution, yields excess hydrogen ions; acids lower the pH. Examples of acids or acidic substances are hydrochloric acid (HCl), sodium acid pyrophosphate (SAPP), and sulfuric acid (H_2SO_4).

Acidity: The relative acid strength of liquid as measured by pH; a compounds having a pH value below 7.

Active system: The volume of drilling fluid being circulated to drill a hole; consists of the volume of drilling fluid in the wellbore plus the volume of drilling fluid in the surface tanks through which the fluid circulates.

Adhesion: The force that holds unlike molecules together.

Adsorb: see Adsorption.

Adsorbed chemical: The molecular species that adhere to the surfaces of solid particles that cannot be removed by draining, even with centrifugal force.

Adsorption: A surface phenomenon exhibited by a solid (adsorbent) to hold or concentrate gases, liquids, or dissolved substances (adsorptive) upon its surface; for example, water held to the outside surface of hydrated bentonite, is adsorbed water.

Aerated fluid: Drilling fluid to which air or gas has been deliberately added to lighten the fluid column.

Aeration: The technique of injecting air or gas in varying amounts into a drilling fluid to reduce the hydrostatic head; also, the inadvertent mechanical incorporation and dispersion of air or gas into a drilling fluid.

Agglomeration: A group of two or more individual particles held together by strong forces; typically, agglomerates are stable to normal stirring, shaking, or handling as powder or a suspension.

Aggregate: To gather together, to clump together; a flocculated drilling fluid will aggregate if flocculent is added.

Aggregation: Formation of aggregates; in drilling fluids, aggregation results in the stacking of the clay platelets face to face and, consequently, the viscosity and gel strength of the fluid decreases.

Agitation: The process of rapidly mixing a slurry within a tank to obtain and maintain a uniform mixture.

Agitator: A mechanically driven impeller used to stir the drilling fluid to assist in the suspension of solids, blending of additives, and maintenance of uniform consistency.

Air cutting: The inadvertent mechanical incorporation and dispersion of air into a drilling fluid system.

Airlock: A condition causing a centrifugal pump to stop pumping because of a large bubble of air or gas in the center of the pump impeller which prevents the liquid from entering the pump suction.

Alkali: Any compound having pH properties higher than the neutral state, i.e., pH >7.

Alkalinity: The combining power of a base measured by the maximum number of equivalents an acid with which it can react to form a salt; in water analyses, alkalinity typically represents the carbonates, bicarbonates, hydroxides, and occasionally the borate, silicates, and phosphates in the water.

Alum: Aluminum sulfate, $Al_2(SO_4)_3$ which is a common inorganic coagulant.

Aluminum stearate: An aluminum salt of stearic acid used as a defoaming agents.

Amorphous: The property of a solid substance that does not crystallize and is without any definite characteristic shape.

Anhydrite: A mineral compound, $CaSO_4$, that is often encountered while drilling; may occur as thin stringers or massive formations.

Anhydrous: Without water.

Aniline point: The lowest temperature at which equal volumes of freshly distilled aniline and an oil sample that is being tested are completely miscible; gives an indication of the molecular characteristics (paraffinic, naphthenic, asphaltic, aromatic) of the oil.

Anion: A negatively charged atom or radical, such as chloride (Cl^-), hydroxide (OH^-), and sulfate (SO^{4-}) in solution of an electrolyte; anions move toward the anode (positive electrode) under the influence of an electrical potential.

Annular pressure loss: The pressure on the annulus required to pump the drilling fluid from the bottom of the hole to the top of the hole in the annular space.

Annular velocity: The velocity of a fluid moving in the annulus, usually expressed in ft/min or m/min.

Annulus (annular space): The space between the drill string and the wall of the wellbore or the inside surface of the casing.

API fluid loss: This fluid loss is measured under ambient conditions – usually room temperature and 100 psi differential pressure.

API gravity: The gravity (weight per unit volume) of crude oil or other related fluids as measured by a system recommended by the American Petroleum Institute (API).

Apparent viscosity: The apparent viscosity in centipoise, as determined by the direct-indicating viscometer is equal to one-half the 600-rpm reading.

Axial flow: Flow from a mechanical agitator in which the fluid first moves along the axis of the impeller shaft (usually down toward the bottom of a tank) and then away from the impeller.

Backpressure: The frictional or blocking pressure opposing fluid flow in a conduit.

Back tank: The compartment on a shale shaker that receives drilling fluid from the flowline.

Backing plate: The plate attached to the back of screen cloth(s) for support.

Baffles: Plates or obstructions built into a compartment to change the direction of fluid flow.

Balanced design hydrocyclone: A hydrocyclone that has the lower apex adjusted to the diameter of the cylinder of air formed within the cone by the cyclonic forces of drilling fluid spinning within cone.

Barium sulfate ($BaSO_4$): used for increasing the density of drilling fluids; barite is the naturally-occurring mineral):

Barrel (bbl): A volumetric unit of measure used in the petroleum industry consisting of 42 U.S. gallons.

Barrel equivalent: One gram of material in 350 ml of fluid is equivalent to a concentration of 1 lb of that material in an oilfield barrel of fluid.

Base: A compound of a metal, or a metal-like group, with hydrogen and oxygen in the proportions that form an OH^- radical, when ionized in an aqueous solution, yielding excess hydroxyl ions; examples of bases are sodium hydroxide (caustic soda, NaOH) and calcium hydroxide [$Ca(OH)_2$].

Base exchange: The replacement of the cations associated with the surface of a clay particle by another species of cation, for example, the substitution of sodium cations by calcium cations on the surface of a clay particle.

Basicity: pH value above 7; the ability to neutralize or accept protons from acids.

Bernoulli Principle: The principle demonstrates that the sum of pressure and velocity through or over a device represents equal quantities, neglecting the effects of losses due to friction and/or increases by adding energy with external devices such as pumps.

Bicarb: Sodium bicarbonate, $NaHCO_3$.

Bingham model: A mathematical description that relates shear stress to shear rate in a linear manner; the model requires only two constants (plastic viscosity and yield point) and is the simplest rheological model possible to describe a non-Newtonian liquid.

Blooie line: The flowline for air or gas drilling.

Blowout: An uncontrolled escape of drilling fluid, gas, oil, or water from the well; caused by the formation pressure being greater than the hydrostatic head of the fluid being circulated in the well bore.

Bottom flooding: The behavior of a hydrocyclone when the underflow discharges whole drilling fluid rather than separated solids.
Brackish water: Water containing low concentrations of any soluble salts.
Break circulation: Commencement of movement of the drilling fluid after it has been quiescent in a borehole.
Brine: Water containing a high concentration of common salts such as sodium chloride, calcium chloride, magnesium chloride, calcium bromide, and zinc bromide.
Buffer: Any substance or combination of substances that, when dissolved in water, produces a solution that resists a change in its hydrogen ion concentration upon the addition of an acid or base.
Cable tool drilling: A method of drilling a well by allowing a weighted bit (or chisel) at the bottom of a cable to fall against the formation being penetrated.
Calcium: One of the alkaline earth elements with a valence of 2 and an atomic weight of 40; a component of lime, gypsum, and limestone; calcium compounds are a common cause of the hardness of water.
Calcium carbonate: $CaCO_3$. An acid soluble calcium salt sometimes used as a weighting material (such as limestone and oyster shell) in specialized drilling fluids.
Calcium chloride: $CaCl_2$. A very soluble calcium salt sometimes added to drilling fluids to impart special inhibitive properties, but used primarily to increase the density of the liquid phase (water) in completion fluids and as an inhibitor to the water phase of invert oil emulsion drilling fluids.
Calcium hydroxide: $Ca(OH)_2$. The active ingredient of slaked lime; also, the main constituent in cement (when wet) and is often referred to as "slaked lime" in field terminology.
Calcium sulfate: Anhydrite, $CaSO_4$, plaster of Paris, $CaSO_4.12H_2O$, and gypsum, $CaSO_4.2H_2O$; occurs in drilling fluids as a contaminant or may be added as a commercial product to certain drilling fluids to impart special inhibitive properties to the fluid.
Calcium-treated drilling fluids: Drilling fluids to which quantities of soluble calcium compounds have been added or allowed to remain from the formation drilled to impart special inhibitive properties to the drilling fluid.
Capacity: The maximum volume flow rate at which a solids-control device is designed to operate without detriment to separation.
Cascade: Gravity-induced flow of fluid from one unit to another.
Casing: Steel pipe placed in an oil or gas well to prevent the wall of the wellbore in a drilled interval from caving in, as well as to prevent movement of fluids from one formation to another.

Cation: The positively charged particle in the solution of an electrolyte, which, under the influence of an electrical potential, moves toward the cathode (negative electrode); examples are Na^+, H^+, NH_4^+, Ca^{2+}, Mg^{2+}, and Al^{3+}.

Cation exchange capacity: The total amount of cations adsorbed on the basal surfaces or broken bond edges of a clay sample, expressed in milliequivalents per 100 grams of dry clay.

Caustic: Sodium hydroxide, NaOH.

Caustic soda: Sodium hydroxide, NaOH; an alkali.

Cavitation: The formation and collapse of low-pressure bubbles in a liquid; cavitation occurs in centrifugal pumps when the pressure within the impeller chamber decreases below the vapor pressure of the liquid.

Cement: A mixture of calcium aluminate derivatives and calcium silicate derivatives made by combining lime and clay while heating; slaked cement contains about 62.5% calcium hydroxide, which can cause a major problem when cement contaminates drilling fluid; used to hold the casing in place and to prevent fluid migration between subsurface formations and it is necessary to know the numbers of sacks of cement required.

Cement plug: The cement plug is part of the well completion process in which a plug of cement slurry placed in the wellbore. Cement plugs are also used for a variety of applications including hydraulic isolation, provision of a secure platform, and in window-milling operations for sidetracking a new wellbore.

Cement sack: The sack, which is a unit of measure for Portland cement in the United States, refers the amount of cement that occupies a bulk volume of 1.0 ft^3. For most Portland cement, including API classes of cement, a sack weighs 94 pounds.

Centipoise (cP): Unit of viscosity equal to 0.01 Poise; the poise equals 1 dyne-second per square centimeter; as an example, the viscosity of water at 20°C (68°F) is 1.005 cP (1 cP = 0.000672 lb/ft sec).

Chemical treatment: The addition of chemicals (such as caustic or thinners) to the drilling fluid to adjust the drilling fluid properties.

Clay: A soft, variously colored earth, commonly hydrous silicates of alumina, formed by the decomposition of feldspar and other aluminum silicates; insoluble in water but disperse under hydration, grinding, or velocity effects; shearing forces break down the clay particles to sizes varying from submicron particles to particles 100 microns or larger.

Clay-size particle: a solid particle having an equivalent spherical diameter less than 2 microns.

Closed loop mud systems: A drilling-fluid processing system that minimizes the liquid discard.

Closed loop systems (pressurized): A system in which formation fluid is contained in tanks and not exposed to the atmosphere until it is sent to the flare line or to the holding tank.

Coagulation: The destabilization and initial aggregation of colloidal and finely divided suspended matter by the addition of a floc-forming agent.

Coalescence: The change from a liquid to a thickened curd-like state by chemical reaction; the combination of globules in an emulsion caused by molecular attraction of the surfaces.

Coarse solids: Solids larger than 2000 microns in diameter.

Connate water: Water trapped within sedimentary deposits, particularly as hydrocarbons displaced most of the water from a reservoir.

Contact time: The time required for a specific fluid to pass a designated depth or point in the annulus during pumping operations; normally used as a design criterion for mud removal in turbulent flow.

Contamination: In a drilling fluid, the presence of any material that may tend to harm the desired properties of the drilling fluid.

Continuous phase: The fluid phase that surrounds the dispersed phase; also, the fluid phase of a drilling fluid, either water, oil, or synthetic oil – the dispersed (non-continuous) phase may be solid or liquid.

Conventional drilling fluid: A drilling fluid containing essentially clay and water; also called conventional mud.

Created fractures: Induced fractures by means of hydraulic or mechanical pressure exerted on the formation by the drill string and/or circulating fluid.

Critical velocity: That velocity at the transitional point between laminar and turbulent types of fluid flow; this point occurs in the transitional range of Reynolds numbers between approximately 2000 to 3000.

Cut point: Cut point curves are developed by dividing the mass of solids in a certain size range removed by the total mass of solids in that size range that enters the separation device.

Cuttings: The pieces of formation dislodged by the bit and brought to the surface in the drilling fluid.

Cyclone: A device for the separation of solid particles from a drilling fluid; the most common cyclones used for solids separation are a desander or desilter.

Darcy: A unit of permeability; for example, a porous medium has a permeability of 1 Darcy when a pressure of 1 atm on a sample 1 cm long and 1 sq cm in cross section will force a liquid of 1 cP viscosity through the sample at the rate of 1 cc per sec.

De-duster: A tank at the end of the blooie line in air or gas drilling in which water is injected to settle the dust caused by drilling.

Deflocculant: A chemical that promotes deflocculation or inhibits flocculation.

Defoamer: Any substance used to reduce or eliminate foam by reducing the surface tension of a liquid.

Degasser: A device that removes entrained gas from a drilling fluid, especially the very small bubbles that do not float readily in viscous drilling fluid.

Dehydration: Removal of free or combined water from a compound or mixture.

Deliquescence: The liquification of a solid substance due to the solution of the solid by absorption of moisture from the air, for example, calcium chloride deliquesces in humid air.

Density: Mass per unit volume expressed in pounds per gallon (ppg), grams per cubic (g/cc), or pounds per cubic ft (lb/cu ft); drilling-fluid density is commonly referred to as mud weight.

Desand: To remove most API sand (>74 microns) from drilling fluid.

Desander: A hydrocyclone larger that can remove a very high proportion of solids larger than 74 micrometers.

Desilt: To remove most silt particles greater than 15 to 20 microns from an unweighted fluid.

Desilter: A hydrocyclone which can remove a large fraction of solids larger than 15 to 20 microns.

Destabilization: A condition in which colloidal particles no longer remain separate and discrete, but contact and agglomerate with other particles.

Diatomaceous earth: A very porous natural earth compound composed of siliceous skeletons; sometimes used for controlling lost circulation and seepage losses and as an additive to cement.

Differential pressure: The difference in pressure between two points; typically, the difference in pressure at a given point in the well bore between the hydrostatic pressure of the drilling-fluid column and the formation pressure; the differential pressure can be positive, zero, or negative with respect to the formation pressure.

Differential sticking: Differential sticking is a condition whereby the drill string cannot be moved (rotated or reciprocated) along the axis of the wellbore and typically occurs when high-contact forces caused by low reservoir pressure, high wellbore pressure, or both, are exerted over a sufficiently large area of the drillstring; differential sticking is a product of the differential pressure between the wellbore and the reservoir and the area that the differential pressure is acting upon; a relatively low differential pressure (Δp) applied over a large working area can be just as effective in sticking the pipe as can a high differential pressure applied over a small area.

Diffusion: The spreading, scattering, or mixing of material (gas, liquid, or solid).

Dilatant fluid: dilatant or inverted plastic fluid is usually made up of a high concentration of well-dispersed solids that exhibit a nonlinear consistency curve passing through the origin; the apparent viscosity increases instantaneously with increasing shear rate and the yield point, as determined by conventional calculations from the direct-indicating viscometer readings, is negative.

Diluent: Liquid added to dilute (thin) a solution or suspension.

Dilution factor: The ratio of the actual volume of drilling fluid required to drill a specified interval of footage using a solids-removal system versus a calculated volume of drilling fluid required to maintain the same drilled-solids fraction over the same specified interval of footage with no drilled-solids removal.

Dilution rate: The rate, in gpm or bbl/hr, at which fluids and/or premix is added to the circulating system for solids management.

Dilution ratio: The ratio of the volume of dilution liquid to the volume of raw drilling fluid in the feed prior to entering a liquid/solids separator.

Dilution water: Water used for dilution of water-based drilling fluid.

Discharge: Material removed from a system; also called effluent.

Dispersant: A chemical that promotes dispersion of particles in a fluid; a deflocculant is often referred to (inaccurately) as a dispersant.

Dispersed phase: The scattered phase (solid, liquid, or gas) of a dispersion; the particles are finely divided and surrounded by the continuous phase.

Disassociation: The splitting of a compound or element into two or more simple molecules, atoms, or ions.

Dog leg (dogleg): The elbow caused by a sharp change of drilling direction in the well bore.

Dolomite: A mineral that is a mixture of calcium carbonate and magnesium carbonate; $CaCO_3.MgCO_3$.

Drill bit (drillbit): The cutting or boring element at the end of the drill string.

Drilled solids: Formation solids that enter the drilling-fluid system, whether produced by a bit or from the side of the borehole.

Drilled-solids fraction: The average volume fraction of drilled solids maintained in the drilling fluid over a specified interval of footage.

Drilled-solids removal system: All equipment and processes used while drilling a well that remove the solids generated from the hole and carried by the drilling fluid, that is, settling, screening, desanding, desilting, centrifuging, and dumping.

Driller's method: The Driller's Method requires two circulations: during the first circulation, the influx is circulated out with the original mud weight. Constant bottom hole pressure is maintained by holding circulating drill pipe pressure constant through the first circulation. See Kill methods.

Drilling cost: The drilling is a summation of several variable, such as the right to enter and drill on the property owner's land is accomplished by obtaining a lease. The lease is subject to title search and proper recording in much the same way as real estate. Between legal cost for title work and lease bonus wells see costs in excess of $1,000,000 for leasing alone.

Drill stem test (DST): A post-drilling and preproduction test that allows formation fluids to flow into the drill pipe under controlled conditions, to determine whether oil and/or gas in commercial quantities have been encountered in the penetrated formations.

Drill string (drillstring): The column of drill pipe with attached tool joints that transmits fluid and rotational power from the kelly to the drill collars and bit.

Drilling fluid: Term applied to any liquid or slurry pumped down the drill string and up the annulus of a hole to facilitate drilling; also called drilling mud or mud.

Drilling-fluid additive: Any material added to a drilling fluid to achieve a desired effect.

Drilling-fluid analysis: Examination and testing of the drilling fluid to determine its physical and chemical properties and functional ability.

Drilling-fluid cycle time: The time necessary to move a fluid from the kelly bushing to the flowline in a borehole. The cycle, in minutes, equals the barrels of drilling fluid in the hole minus pipe displacement divided by barrels per minute of circulation rate.

$$(\text{Hole}_{bbl} - \text{Pipe Voume}_{bbl}) = \text{Circulation Rate}_{bbl/min}$$

Drilling in: The drilling operation starting at the point of drilling into the producing formation.

Drilling mud: Drilling fluid.

Drilling out: The operation of drilling out of the casing shoe after the cementing of a casing or liner in place; drilling out of the casing is done before further hole is made or completion attempted.

Drilling rate: The rate at which hole depth progresses, expressed in linear units per unit of time (including connections) as feet/minute or feet/hour.

Dry bottom: An adjustment to the underflow opening of a hydrocyclone that causes a dry beach, usually resulting in severe plugging.

Dry plug: The plugging of the underflow opening of a hydrocyclone caused by operating with a dry bottom.

Effective screening area: The portion of a screen surface available for solids separation.

Effluent: A discharge of liquid; generally, a stream of liquid after separation or purification; also called discharge.

Electric logging: Logs run on a wire line to obtain information concerning the porosity, permeability, density, and/or fluid content of the formations drilled; the drilling-fluid characteristics may need to be altered to obtain good logs.

Electrolyte: A substance that dissociates into charged positive and negative ions when in solution or a fused state; an electrolyte will then conduct an electric current and acids, bases, as well as salts are common electrolytes.

Elevation head: The pressure created by a given height of fluid; also called hydrostatic pressure head.

Emulsifier: A chemical substance used to produce a mixing of two liquids that do not solubilize in each other or maintain a stable mixture when agitated in the presence of each other.

Emulsion: A substantially permanent heterogeneous mixture of two or more liquids that do not normally dissolve in each other but are held in a dispersed state, one within the other; the dispersion is accomplished by the combination of mechanical agitation and presence of fine solids and/or emulsifiers.

Encapsulation: The process of totally enclosing electrical parts or circuits with a polymeric material (usually epoxy resin).

End point: Indicates the end of a chemical testing operation when a clear and definite change is observed in the test sample; in titration, this change is frequently a change in color of an indicator or marker added to the solution, or the disappearance of a colored reactant.

Equivalent circulating density (ECD): The effective drilling-fluid weight at any point in the annulus of the well bore during fluid circulation; includes drilling-fluid density, cuttings in the annulus, and annular pressure loss.

Equivalent spherical diameter (ESD): The theoretical dimension usually referred to when the sizes of irregularly shaped particles are discussed; these dimensions can be determined by several methods, such as settling velocity, electrical resistance, and light reflection.

Equivalent weight: The atomic weight or formula weight of an element, compound, or ion divided by its valence; elements entering into chemical combination or chemical reaction always do so in quantities to their equivalent weights; also known as combining weight.

Expected cost: The probability that a certain cost will be incurred.

Expected profit is the probability of receiving a certain profit.

Expected value: The difference between expected profits and expected costs. See also: Future value.

Fault: Geological term denoting a formation break across the trend of a subsurface strata; faults can significantly affect the drilling fluid and casing programs due to possibilities for lost circulation, sloughing hole, or kicks.

Feed (feedstock): A mixture entering a liquid/solids separation device or a reactor.

Feed capacity (feedstock capacity): The maximum volume flow rate at which a solids-control device is designed to operate without detriment to separation efficiency; the capacity will be dependent on particle size, particle concentration, viscosity, and other variables of the feed (feedstock).

Feed chamber: That part of a device that receives the mixture of diluents, drilling fluid, and solids to be separated.

Feed head: The equivalent height, in feet or meters, of a column of fluid at the cyclone feed header.

Feed header: A pipe, tube, or conduit to which two or more hydrocyclones have been connected and from which they receive their feed slurry.

Feed inlet: The opening through which the feed fluid enters a solids separation device. Also known as feed opening.

Feed pressure: The actual gauge pressure measured as near as possible to, and upstream of, the feed inlet of a device.

Filter cake: The suspended solids that are deposited on the filter during the process of filtration.

Filter cake texture: The physical properties of a cake as measured by toughness, slickness, and brittleness.

Filtrate: The liquid that is forced through a porous medium during the filtration process.

Filter press: A device for determining the fluid loss of a drilling fluid.

Filter run: The interval between two successive backwashing operations of a filter.

Filterability: The characteristic of a clear fluid that denote both the ease of filtration and the ability to remove solids while filtering.

Fine solids: Solids 44 to 74 microns in diameter, or sieve size 325 to 200 mesh.

Fishing: Operations on the rig for retrieving sections of pipe, collars, or other obstructive items that are in the wellbore and would interfere with drilling or logging operations.

Flat gel: A condition wherein the gel strength does not increase appreciably with time and is essentially equal to the initial gel strength; opposite of progressive gel.

Floc: Small masses of gelatinous solids formed in a liquid.

Flocculates: A group of aggregates or particles in a suspension formed by electrostatic attraction forces between negative and positive charges.

Flocculating agent: Substances, for example, most electrolytes, a few polysaccharides, certain natural or synthetic polymers, that bring about the thickening of a drilling fluid; in Bingham plastic fluids, the yield point and gel strength increase with flocculation.

Flooding: The effect created when a screen, hydrocyclone, or centrifuge is fed beyond its capacity; flooding may also occur on a screen because of blinding.

Flowback pan: A pan or surface below a screen that causes fluid passing through one screen to flow back to the feed end of a lower screen.

Flow capacity: The rate at which a shaker can process drilling fluid and solids; depends on many variables, including shaker configuration, design and motion, drilling fluid rheology, solids loading, and blinding by near-size particles.

Flow drilling: Drilling in which there is a constant flow of formation fluid.

Flowline: The pipe (usually) or trough that conveys drilling fluid from the rotary nipple to the solids-separation section of the drilling fluid tanks on a drilling rig.

Flow rate: The volume of liquid or slurry moved through a pipe in one unit of time, such as gpm or bbl/min.

Flow streams: With respect to centrifugal separators, all liquids and slurries entering and leaving a machine, such as feed drilling fluid stream plus dilution stream equals overflow stream plus underflow stream.

Fluid: Any substance that will readily assume the shape of the container in which it is placed; includes both liquids and gases; a substance in which the application of every system of stress (other than hydrostatic pressure) will produce a continuously increasing deformation without any relation between time rate of deformation at any instant and the magnitude of stress at the instant.

Fluid flow: The state of dynamics of a fluid in motion as determined by the type of fluid (e.g., Newtonian plastic, pseudoplastic, dilatant), the properties of the fluid such as viscosity and density, the geometry of the system, and the velocity; under a given set of conditions and fluid properties, the fluid flow can be described as plug flow, laminar (called also Newtonian, streamline, parallel, or viscous) flow, or turbulent flow.

Fluid loss: The relative amount of fluid loss (filtrate) through permeable formations or membranes when the drilling fluid is subjected to a pressure differential.

Fluidity: The reciprocal of viscosity; the ease of flowing; the measure of rate with which a fluid is continuously deformed by a shearing stress.

Fluorescence: Instantaneous re-emission of light of a greater wavelength than that of the light originally absorbed.

Flute: A curved metal blade wrapped around a shaft as on a screw conveyor in a centrifuge.

Foam: A two-phase system that is similar to an emulsion, in which the dispersed phase is a gas or air; bubbles (usually air but can be formation gas) floating on the surface of the drilling fluid.

Foaming agent: A substance that produces stable bubbles at the air/liquid interface due to agitation, aeration, or ebullition; in air or gas drilling, foaming agents are added to turn water influx into aerated foam; commonly referred to as "mist drilling".

Foot: Unit of length (12 inches) in British (foot-pound-second) system.

Foot-pound: Unit of work or of mechanical energy, which is the capacity to do work; 1 foot is the work performed by a force of 1 pound acting through a distance of 1 foot; or the work required to lift a 1-pound weight a vertical distance of 1 foot.

Foot valve: A check valve installed at the suction end of a suction line.

Formation: A bed or deposit (stratum) composed throughout of substantially the same type of rock (mineral).

Formation damage: Damage caused by the invasion of the formation by drilling-fluid particles, drilling-fluid filtrates, and/or cement filtrates; can also result from changes in pH and a variety of other conditions; asphaltene constituents deposited from crude oil will also damage some formations.

Formation fluid: The fluid (brine, oil, gas) that exists in the pores of a formation.

Formation sensitivity: The tendency of certain producing formations to adversely react with the drilling and completion process.

Free liquid: The liquid film that can be removed by gravity draining or centrifugal force.

Free-water knockout: A water/gas separator ahead of the flare line.

Freshwater drilling fluid: A drilling fluid in which the liquid phase is freshwater.

Future value: The value of a current asset at a specified date in the future based on an assumed rate of growth over time. See also: Expected value.

Using simple interest, future value = P x I x N where P is the initial investment amount, I is the interest rate and N is the number of years the investment will be held.

For compound interest, the future value = P x [(1 + I)^N – 1] where P is the principal, I = nominal annual interest rate, and N = the number of compounding periods.

Galena: Lead sulfide (PbS); technical grades (specific gravity about 7.0) are used for increasing the density of drilling fluids to points impractical or impossible with barite. Almost entirely used in preparation of kill fluids.

Gas buster: Mud/gas separator.

Gas cut: Gas entrained by a drilling fluid.

Gel: A state of a colloidal suspension in which shearing stresses below a certain finite value fail to produce permanent deformation; the minimum shearing stress that will produce permanent deformation is known as the shear or gel strength of the gel; also, a term used to designate highly colloidal, high-yielding, viscosity-building, commercial clays, such as bentonite and attapulgite.

Gelation: Association of particles forming continuous structures at low shear rates.

Gel cement: Cement having a small to moderate percentage of bentonite added as a filler and/or reducer of the slurry weight; the bentonite may be dry-blended into the mixture or added as a prehydrated slurry.

g Factor: The acceleration of an object relative to the acceleration of gravity.

g Force: The centrifugal force exerted on a mass moving in a circular path.

Gypsum: Calcium sulfate, $CaSO_4$.2H2O, frequently encountered while drilling; may occur as thin stringers or in massive formations.

Head: The height a column of fluid would stand in an open-ended pipe if it was attached to the point of interest; the head at the bottom of a 1000-ft well is 1000 ft, but the pressure would be dependent on the density of the drilling fluid in the well.

Heterogeneous: A substance that consists of more than one phase and is not uniform, such as colloids, emulsions, and has different properties in different parts.

High-gravity solids (HGS): Solids purchased and added to a drilling fluid specifically and solely to increase drilling-fluid density; barite (specific gravity: 4.2) and hematite (specific gravity: 5.05) are the most common additives used for this purpose.

High-pH drilling fluid: A drilling fluid with a pH range above 10.5.

Horsepower: The rate of doing work or of expending mechanical energy; that is, horsepower is work performed per unit of time.

Horsepower-hour: Horsepower-hour (hp-hr) and kilowatt-hour (kW-hr) are units of work.

Horseshoe effect: The U shape formed by the leading edge of drilling fluid moving down a shale shaker screen; the drilling fluid usually tends to pass through the center of a crowned screen faster than it passes through the edges, creating the U shape.

HTHP: High temperature high pressure.

Humic acid: Organic acids of indefinite composition found in naturally occurring lignite; the humic acids are the active constituents that assist in the positive adjustment of drilling-fluid properties.

Hydrate: A substance containing water combined in molecular form (such as $CaSO_4.2H_2O$). A crystalline substance containing water of crystallization.

Hydration: The act of a substance to take up water by means of absorption and/or adsorption; usually results in swelling, dispersion and disintegration into colloidal particles.

Hydrocyclone: A liquid/solids separation device utilizing centrifugal force; the fluid tangentially and spins inside the cone causing the heavier solids to settle to the walls of the cone and move downward until they are discharged at the cone bottom (cone apex).

Hydrocyclone underflow: The discharge stream from a hydrocyclone that contains a higher percentage of solids than does the feed.

Hydrogen ion concentration: A measure of either the acidity or alkalinity of a solution, normally expressed as pH.

Hydrolysis: The reaction of a salt with water to form an acid or base; for example, soda ash (sodium carbonate, Na_2CO_3) hydrolyzes basically, and hydrolysis is responsible for the increase in the pH of water when soda ash is added.

Hydrometer: A floating instrument for determining the specific gravity or density of liquids, solutions, and slurries.

Hydrophile: Any substance, usually in the colloidal state or an emulsion, that is wetted by water; that is, it attracts water or water adheres to it.

Hydrophilic: A property of a substance having an affinity for water or one that is wetted by water.

Hydrophobe: Any substance, usually in the colloidal state, that is not wetted by water.

Hydrophobic: Any substance, usually in the colloidal state or an emulsion, that is not wetted by water; that is, it repels water or water does not adheres to it.

Hydrostatic pressure: The pressure exerted by a fluid at equilibrium at a given point within the fluid, due to the force of gravity. The hydrostatic pressure increases in proportion to depth measured from the surface because of the increasing weight of fluid exerting downward force from above.

Hydrostatic pressure head: The pressure exerted by a column of fluid, usually expressed in pounds per square inch.

Hydroxide: Designation that is given basic compounds containing the hydroxyl (OH^-) radical; when these substances are dissolved in water, the pH of the solution is increased.

Hygroscopic: The property of a substance enabling it to absorb water from the air.

ID: Inside (internal) diameter of a pipe.

Inhibited drilling fluid: A drilling fluid having an aqueous phase with a chemical composition that tends to retard and even prevent (inhibit) appreciable hydration (swelling) or dispersion formation clays and shales through chemical and/or physical reactions.

Inlet: The opening through which the feed mud enters a solids-control device.

Interfacial tension: The force required to break the surface definition between two immiscible liquids; the lower the interfacial tension between the two phases of an emulsion, the greater the ease of emulsification and when the values approach zero, emulsion formation is spontaneous.

Intermediate solid: A particles with a diameter between 250 and 2000 microns.

Interstitial water: Water contained in the interstices or voids of formations.

Invert oil emulsion drilling fluid: A water-in-oil emulsion in which water (sometimes containing sodium or calcium chloride) is the dispersed phase, and diesel oil, crude oil, or some other oil is the phase.

Ions: Molecular condition due to loss or gain of electrons; acids, bases, and salts electrolytes), when dissolved in certain solvents, especially water, are dissociated into electrically charged ions or parts of the molecules; loss of electrons results in positive charges, producing a cation whereas a gain of electrons in the formation of an anion, with negative charge. The valence of an ion is equal to the number of charges borne by the ion.

Jones effect: The net surface tension of all salt solutions first decreases with an increase in concentration, passes through a minimum, and then increases as the concentration is raised; the initial decrease is the Jones effect.

Kelly: A heavy square or hexagonal pipe that passes through rollers in a bushing on the drill floor to transmit rotational torque to the drill string.

Keyseat (key seat): A section of a borehole, usually of abnormal deviation and relatively soft formation, that has been eroded or worn by drill pipe to a size smaller than the tool joints or collars of the drill string; the keyhole-type configuration resists passage of the shoulders of the pipe upset (box) configurations when pulling out of the hole.

Kick: A situation caused when the annular hydrostatic pressure in a drilling well temporarily (and usually relatively suddenly) becomes less than the formation, or pore, pressure in a permeable downhole section.

Kill fluid: A fluid built with a specific density aimed at controlling a kick or blowout.

Kill line: A line connected to the annulus below the blowout preventers for pumping into the annulus while the preventers are closed.

Killing a well: Bringing a well kick under control; also, the procedure of circulating a fluid into a well to overbalance formation fluid pressure after the bottom-hole pressure has been less than formation fluid pressure.

Kill methods: The two widely used constant bottomhole circulating methods are the Driller's Method and the Wait and Weight (W&W) Method. Well control experts are often strongly opinionated on selecting the better method to circulate an influx out of the wellbore. The purpose of this article is to highlight the major advantages and disadvantages of the two methods. The basic principle of both methods is to keep bottomhole pressure (BHP) constant at or, preferably, slightly above the formation pressure. See Driller's method and the W&W method

Kinematic viscosity: The kinematic viscosity of a fluid is the ratio of the viscosity (e.g., cP in g/cm-sec) to the density (e.g., g/cc) using consistent units.

Leak-off pressure: Typically leak-off occurs when the fracture area is constant during shut-in and the leak-off occurs through a homogeneous rock matrix.

Lime: $Ca(OH)_2$: A commercial form of calcium hydroxide.

Lime-treated drilling fluids: Commonly referred to as "lime-based muds"; these high-pH systems contain most of the conventional freshwater drilling-fluid additives to which slaked lime has been added to impart special inhibition properties.

Limestone: Calcium carbonate, $CaCO_3$.

Line sizing: Ensuring that the fluid velocity through all piping within the surface system has the proper flow and pipe diameter combination to prevent solids from settling and pipe from eroding.

Lipophile: Any substance, usually in the colloidal state or an emulsion, that is wetted by oil; that is, it attracts oil or oil adheres to it.

Lipophilic: A property of a substance having an affinity for oil or one that is wetted by oil.

Live oil: Crude oil that contains gas and distillates and has not been stabilized or weathered; can cause gas cutting when added to drilling fluid and is a potential fire hazard.

Lost circulation: The result of drilling fluid escaping into a formation, usually in fractures, cavernous, fissured, or coarsely permeable beds, evidenced by the complete or partial failure of the drilling fluid to return to the surface as it is being circulated in the hole.

Lost circulation additives: Materials added to the drilling fluid to gain control of or prevent the loss of circulation; these materials are added in varying amounts and are classified as fibrous, flake, or granular.

Low-gravity solids: Salts, drilled solids of every size, commercial colloids, lost circulation materials; that is, all solids in drilling fluid, except barite or other commercial weighting materials; salt (sodium chloride, NaCl) is considered a low-specific gravity solid.

Low-silt drilling fluid: An unweighted drilling fluid that has all the sand and a high proportion of the silts removed and has a substantial content of bentonite or other water–loss–reducing clays.

Low-solids drilling fluids: A drilling fluid that has polymers, such as ceramic matrix compound (CMC) or xanthan gum (XC) polymer, partially or wholly substituted for commercial or natural clay minerals.

Low-solids non-dispersed (LSND) drilling fluids: A drilling fluid to which polymers have been added to simultaneously extend and flocculate bentonite drilled solids.

Low-yield clay: Commercial clay chiefly of the calcium montmorillonite type having a yield of approximately 15 to 30 barrels per ton.

Lyophilic: Having an affinity for the suspending medium, such as bentonite in water.

Lyophilic colloid: A colloid that is not easily precipitated from a solution and is readily dispersible after precipitation by addition of a solvent.

Lyophobic colloid: A colloid that is readily precipitated from a solution and cannot be redispersed by addition of the solution.

Manifold: A length of pipe with multiple connections for collecting or distributing fluid; also, a piping arrangement through which liquids, solids, or slurries from one or more sources can be fed to or discharged from a solids-separation device.

Marsh funnel: An instrument used in determining the Marsh funnel viscosity.

Marsh funnel viscosity: Commonly called "funnel viscosity"; reported as the time, in seconds, required for 1 quart of fluid to flow through an API standardized funnel – in some laboratories the efflux quantity is 1000 cc.

Martin's radii: The distance from the centroid of an object to its outer boundary; the direction of this measurement is specified by the azimuth orientation of the line from horizontal.

Median cut: The median cut is the particle size that reports 50% of the weight to the overflow 50% of the weight to the underflow; often identified as the D50 point.

Medium (solids): Particles with a diameter between 74 and 250 microns.

Meniscus: The curved upper surface of a liquid column, concave when the containing walls are wetted by the liquid and convex when they are not wetted.

Mica: A naturally occurring alkali aluminum silicate mineral flake material of various sizes used in controlling lost circulation.

Micelle: An organic and/or inorganic molecular aggregate occurring in colloidal solutions; may appear as chains of individual structural units chemically joined to one another and deposited side by side to form bundles.

Micron: A unit of length equal to one-thousandth of a millimeter; used to specify particle sizes in drilling fluids and solids control discussions (25,400 microns = 1 inch).

Mil: A unit of length equal to 1/1000 inch.

Milliliter: A metric system unit for the measurement of volume – 1/1000th of a liter; in drilling-fluid analyses, this term is used interchangeably with cubic centimeter (cc); one quart is equal to approximately 946 ml.

Mist drilling: A method of rotary drilling whereby water and/or oil is dispersed in air and/or gas as the drilling fluid.

Montmorillonite: A clay mineral commonly used as an additive to drilling mud; sodium montmorillonite is the main constituent of bentonite.

Mud: Drilling fluid, which is the preferred term.

Mud balance: A beam-type balance used in determining drilling-fluid density (mud weight); consists primarily of a base, a graduated beam with constant volume cup, lid, rider, knife-edge, and counterweight.

Mud gun: A submerged nozzle used to stir the drilling fluid with a high-velocity stream.

Mud inhibitor: Additives such as salt, lime, lignosulfonate, and calcium sulfate that prevent clay dispersion.

Mud logging: A process that helps determine the presence or absence of oil or gas in the various formations penetrated by the drill bit, and assists with a variety of indicators that assist drilling operations.

Mud weight: A measurement of density of a slurry usually reported in lb/gal, lb/cu ft, psi/1000 ft or specific gravity.

Mud/gas separator: A vessel into which the choke line discharges when a "kick" is being taken; gas is separated in the vessel as the drilling fluid flows over baffle plates.

Mudding off: A condition promoting reduced production caused by the penetrating, sealing or plastering effect of a drilling fluid.

Mudding up: Process of mixing drilling fluid additives to a simple, native clay water slurry to achieve some properties not possible with the previous fluid.

MW: Abbreviation for mud weight.

Neat cement: A slurry composed only of Portland cement and water.

Negative deck angle: The angle of adjustment to a screen deck that causes the screened solids to travel downhill to reach the discharge end of the screen surface; the downhill travel decreases the fluid throughput of a screen but usually lengthens the life of a screen.

Neutralization: A reaction in which the hydrogen ion of an acid and the hydroxyl ion of a base unite to form water, the other ionic product being a salt.

Newtonian fluid: A fluid in which the shear force is directly proportional to the shear rate and the fluids will immediately begin to move when a pressure or force more than zero psi is applied; examples of Newtonian fluids are water, diesel oil, and glycerine.

Oil-based drilling fluid (oil-based mud): The term is applied to a drilling fluid in which oil is the continuous phase and water the dispersed phase; typically contains from 1 to 5% water emulsified into the system with lime and emulsifiers.

Oil breakout: Oil that has risen to the surface of a drilling fluid; this oil had been previously emulsified in the drilling fluid or may derive from oil-bearing formations that have been penetrated.

Oil wet: A surface on which oil easily spreads; if the contact angle of an oil droplet on a surface is less than 90°, the surface is oil wet.

Oil-in-water emulsion drilling fluid: Any conventional or special water-base drilling fluid to which oil has been added; typically, the oil content is usually kept between 3 and 7% v/v and seldom over 10% v/v.

Overflow: The discharge stream from a centrifugal separation that normally contains a higher percentage of liquids than does the feed.

Overflow header: A pipe into which two or more hydrocyclones discharge their overflow.

Packer fluid: A fluid placed in the annulus between the tubing and casing above a packer; the hydrostatic pressure of the packer fluid is utilized to reduce the pressure differentials between the formation and the inside of the casing and across the packer.

Particle: A discrete unit of solid material that may consist of a single grain or of any number of grains stuck together.

Particle size: Particle diameter expressed in microns.

Particle size distribution: The classification of solid particles into each of the various size ranges as a percentage of the total solids of all sizes in a sample.

Parts per million: The unit weight of solute per million unit-weights of solution (solute plus solvent), corresponding to weight percentage.

Pay zone: A formation that contains oil and/or gas in commercial quantities.

Penetration rate: The rate at which the drill bit penetrates the formation, usually expressed in feet per hour or meters per hour.

pH: The negative logarithm of the hydrogen ion concentration in gram ionic weights per liter; the pH range is numbered from 0 to 14, with 7 being neutral; an index of the acidity (below 7) or alkalinity (above 7) of the fluid.

Pit gain: Fluids entering the wellbore displace an equal volume of mud at the flowline, resulting in pit gain.

Plastic viscosity: A measure of the internal resistance to fluid flow attributable to the concentration, type, and size of solids present in a fluid and the viscosity of the continuous phase; the value, expressed in centipoise, is proportional to the slope of the shear stress/shear rate curve determined in the region of laminar flow for materials whose properties are described by Bingham's law of plastic flow.

Plug flow: The movement of material as a unit without shearing within the mass; typically the flow exhibited by a plastic fluid after overcoming the initial force required to produce flow.

Pool: The reservoir or pond of fluid, or slurry, formed inside the wall of hydrocyclones and centrifuges and in which classification or separation of solids occurs due to the settling effect of centrifugal force; also, the reservoir or pond of fluid that can form on the feedstock end of an uphill shaker basket, a shaker basket with positive deck angle.

Porosity: The volume of void space in a formation rock usually expressed as percentage of void volume per bulk volume.

Pound equivalent: A laboratory unit used in pilot testing; one gram of a material added to 350 ml of fluid is equivalent to 1 lb of material added to one barrel.

ppm: Parts per million.

Prehydration tank: A tank used to hydrate materials (such as bentonite and polymers) that require a long time (hours to days) to hydrate fully and disperse before being added to the drilling fluid.

Premix system: A compartment used to mix materials (such as bentonite) that require time to hydrate or disperse fully before they are added to the drilling fluid.

Pressure head: Pressure within a system equal to the pressure exerted by an equivalent height of fluid (expressed in feet or meters).

Pressure loss: The pressure lost in a pipeline or annulus due to the velocity of the liquid in the pipeline, the properties of the fluid, the condition of the pipe wall, and the configuration of the pipe.

Pressure surge: A sudden, usually brief increase in pressure that can occur when pipe or casing is run into a borehole too rapidly or the drill string is set in the slips too quickly - an increase in the hydrostatic pressure results due to pressure surge which may be great enough to create lost circulation.

Price elasticity: Price elasticity is the ratio of the percentage of change of wells and footage drilled to the percentage change in the crude price and describes the degree of responsiveness of the rig in demand or rig in supply to the change in the crude price; used to measure the effect of economic variables such as demand or supply of rigs or wells drilled with respect to change in the crude oil price and enables the company to discover the sensitivity one variable is with the other one, and it is also independent of units of measurement;

Purging: The process of supplying an enclosure with a protective gas at a sufficient flow and positive pressure to reduce the concentration of any flammable gas or vapor initially present to an acceptable level.

Quicklime: Calcium oxide, CaO. Used in certain oil-based drilling fluids to neutralize the organic acid(s).

Radial flow: Flow of a fluid outwardly in a 360° pattern which describes the flow from a mechanical agitator in which fluid moves away from the axis of the impeller shaft (usually horizontally toward a mud tank wall).

Rate of penetration: The rate at which the drill bit penetrates the formation, expressed in lineal units of feet/minute.

Rate of shear: The change in velocity between two parallel layers divided by the distance between the layers. Shear rate has the units of reciprocal seconds (sec^{-1}).

Raw drilling fluid: Drilling fluid, before dilution, that is to be processed by solids-removal equipment.

Retention time: The time any given particle of material is retained in a region, for example, the time a particle is on a screening surface, within a hydrocyclone, or within the bowl of a centrifuge.

Retort: An instrument used to distill oil, water, and other volatile material in a drilling fluid to determine oil, water, salt, and total solids contents in volume percentage.

Reverse circulation: The method by which the normal flow of a drilling fluid is reversed by circulating down the annulus, then up and out the drill string.

Reynolds number: A dimensionless number, R_e, that occurs in the theory of fluid dynamics. The Reynolds number for a fluid flowing through a cylindrical conductor is determined by the equation.

$$Re = DV\rho/\mu$$

D is the diameter, V is the velocity, ρ is the density, and μ is the viscosity; the number can be used to indicate the type of fluid flow – the transitional range occurs approximately from 2000 to 3000 and below 2000, the flow is laminar but above 3000, the flow is turbulent.

Rig pump: The reciprocating, positive displacement, high-pressure pump on a drilling rig used to circulate the hole.

Rig shaker: Slang term for a shale shaker.

ROP: Rate of penetration.

Rotary drilling: The method of drilling wells in which a drill bit attached to a drill string is rotated on the formation to be drilled; a fluid is circulated through the drill pipe to remove cuttings from the bottom of the hole, bring cuttings to the surface, and perform other functions.

Rotary mud separator (RMS): A centrifuge consisting of a perforated cylinder rotating inside of an outer cylinder housing.

Sack: A unit of measure for Portland cement in the United States and generally refers to the amount of cement that occupies a bulk volume of 1.0 ft^3. For most Portland cement, including API classes of cement, a sack weighs 94 pounds.

Centipoise (cP): Unit of viscosity equal to 0.01 Poise; the poise equals 1 dyne-second.

Salt: A class of compounds formed when the hydrogen of an acid is partially or wholly replaced by a metal or a metallic radical.

Saltwater drilling fluid: A water-based drilling fluid whose external liquid phase contains sodium chloride or calcium chloride.

Sand: Geological term; also used in the particle-size classification system for solids larger than 74 microns; a loose, granular material resulting from the disintegration of rocks with a high silica content.

Separator: A tank in which mixed water, oil, and gas are separated by gravity or enhanced force.

Separator (open/atmospheric): A separator for drilling fluid/formation fluid that is open to atmospheric pressure.

Separator (closed/pressurized): A separator for drilling fluid/formation fluid that is closed and pressurized.

Settling velocity: The velocity a particle achieves in a fluid when gravity forces equal friction forces of the moving particle, that is, when the particle achieves its maximum velocity.

Shear rate: The change of velocity with respect to the distance perpendicular to the velocity changes.

Shear stress: The force per unit of an area parallel to the force that tends to slide one surface past another.

Sieve analysis: The mass classification of solid particles passing through or retained on a sequence of screens of increasing mesh count.

Slip: The difference between synchronous speed and operating speed compared with synchronous speed, expressed as a percentage; if expressed in rpm, slip is the difference between synchronous speed and operating speed.

Slip velocity: The slip velocity arises because of the difference between the average velocities of two different fluids flowing together in a pipe. In vertical ascending flow, the lighter fluid flows faster than the heavier fluid. The slip velocity depends mainly on the difference in density between the two fluids, and their respective hold up.

Sloughing: A situation in which portions of a formation fall away from the walls of a hole, because of incompetent unconsolidated formations, tectonic stresses, high angle of repose, wetting along internal bedding planes, or swelling of formations.

Slurry: A mixture or suspension of solid particles in one or more liquids.

Sodium bicarbonate: $NaHCO_3$. A material used extensively for treating cement contamination and occasionally other calcium contamination of drilling fluids: It is the half-neutralized salt of carbonic acid.

Sodium chloride: NaCl. Commonly known as salt; may be present in the drilling fluid as a contaminant or may be added purposely for inhibition.

Sodium hydroxide: NaOH. Commonly referred to as "caustic" or "caustic soda"; a chemical used primarily to raise pH.

Sodium silicate drilling fluids: Special class of inhibited chemical drilling fluid using sodium silicate, saltwater, and clay.

Solute: A substance that is dissolved in another (the solvent).

Solution: A mixture of two or more components that form a homogeneous single phase. An example of a solution is salt dissolved in water.

Solvent: Liquid used to dissolve a substance (the solute).

Specific gravity (SG): The weight of a specific volume of a liquid, solid, or slurry in reference to the weight of an equal volume of water at a reference temperature of 3.89°C (water has a density of 1.0 g/cc at this temperature).

Specific heat capacity: The number of calories required to raise 1 g of a substance one degree Celsius.

Spotting fluid: The term used for a small volume or pill of fluid placed in a wellbore annulus to free differentially stuck pipe. Oil-base drilling mud is the traditional stuck-pipe spotting fluid. Speed in mixing and placing the spot is of primary importance to successfully freeing pipe.

Spud mud: The drilling fluid used when drilling starts at the surface, often a thick bentonite-lime slurry.

Spudding in: The initiating of the drilling operations in the first top-hole section of a new well.

Squirrel-cage motor: An induction motor that gets its name from the rotor assembly that looks like a squirrel cage, typical of those used earlier in the twentieth century.

Stacking a rig: Storing a drilling rig upon completion of a job when the rig is to be withdrawn from service for a period of time.

Stiff foam: A foam in which a bentonite or long-chain polymer has been added.

Stoke's law: The law that states that the terminal settling velocity of a spherical particle is proportional to the square of the particle diameter, the acceleration of gravity, and the density difference between the density of the particle and the density of the liquid medium; the terminal settling velocity is inversely proportional to the viscosity of the liquid medium.

Stuck point: During drilling operations, a pipe is considered stuck if it cannot be freed from the hole without damaging the pipe, and without exceeding the maximum allowed hook load of the drilling rig.

Sump: A disposal compartment or earthen pit for holding discarded liquids and solids.

Surface tension: Generally, the cohesive forces acting on surface molecules at the interface between a liquid and its own vapor; this force appears as a tensile force per unit length along the interface surface and is usually expressed in units of dynes per centimeter.

Surfactant: Material that tends to concentrate at an interface of an emulsion or a solid/liquid interface; used in drilling fluids to control, for example, the degree of emulsification, aggregation, dispersion, interfacial tension, foaming, defoaming, and wetting

Surge: An increase of the pressure in a well bore that is caused by lowering tubulars.

Tannic acid: The active ingredient of quebracho and other quebracho substitutes such as mangrove bark, chestnut extract, and hemlock.

Temperature survey: An operation to determine temperatures at various depths in the well bore; the survey is used to find the location of inflows of water into the borehole or where proper cementing of the casing has taken place.

Thermal decomposition: Chemical breakdown of a compound or substance by temperature into simple substances or into its constituent elements.

Thinner: Any of the various organic agents (such as tannin derivatives, lignin derivatives, and lignosulfonate derivatives) and inorganic agents (pyrophosphate derivatives and tetraphosphate derivatives) that are added to a water-based drilling fluid to reduce the low-shear-rate viscosity and/or thixotropic properties by deflocculation.

Thixotropy: The ability of a fluid to develop gel strength with time; the property of a fluid at rest that causes it to build up a rigid or semi-rigid gel structure if allowed to remain at rest – the change is reversible and the fluid can be returned to a liquid state by mechanical agitation.

Tool joint: A drill-pipe coupler consisting of a threaded pin and a box of various designs and sizes.

Torque: The turning effort caused by a force acting normal to the radius at a specified distance from the axis of rotation; torque is expressed in pound-feet (pounds at a radius of one foot).

Total depth (TD): The greatest depth reached by the drill bit in a well.

Transport velocity: In rotary drilling operations, both the fluid and the rock fragments are moving in the annulus and the situation is complicated by the fact that the fluid velocity varies from zero at the wall to a maximum at a point between the pipe outer wall and the wellbore wall. In addition, the rotation of the drill pipe imparts centrifugal force on the rock fragments, which affects their relative location in the annulus. In practice, either the flow rate or effective viscosity of the fluid is increased, if problems related to inefficient cuttings removal are encountered and the result is a natural tendency toward thick mud and high annular velocity. However, increasing the mud viscosity or flow rate can be detrimental to the cleaning action beneath the bit, and cause a reduction in the penetration rate.

Turbulent flow: Fluid flow in which the velocity varies in magnitude and the direction of flow; pursues erratic and continually varying courses.

Twist-off: The severing or failure of a joint of drill pipe caused by excessive torque.

Ultra-fine solid: A particle with a diameter between 2 and 44 microns.

Unweighted drilling fluid: A drilling fluid that does not contain commercial suspended solids added to increase the density of the drilling fluid.

Valence: A number representing the combining power of an atom, that is, the number of electrons lost, gained, or shared by an atom in a compound.

Viscosity: The ratio of shear stress to shear rate in a fluid; if the shear stress is measured in dynes/cm^2 and the shear rate in reciprocal seconds, the ratio is the viscosity, in Poise; the internal resistance offered by a fluid to flow.

Volatile matter: Normally gaseous products given off by a substance, such as gas breaking out of live crude oil that has been added to a drilling fluid.

W&W method: The W&W method (wait and weight method, sometimes called the engineer's method because it involves more calculations than the Driller's method) involves only one circulation; the influx is circulated out, and the kill mud is pumped in one circulation; while pumping kill mud from surface to bit, a drill pipe pressure schedule has to be calculated and followed. The drill pipe pressure is held constant thereafter until kill mud is observed returning to the surface. See Kill methods.

Wall cake: The solid material deposited along the wall of the hole resulting from filtration of the fluid part of the drilling fluid into the formation.

Water-based drilling fluid: Common, conventional drilling fluid; water is the suspending medium for solids and is the continuous phase, whether or not oil is present.

Water wet: A surface on which water easily spreads; if the contact angle of a water droplet on a surface is less than 90°, the surface is water wet.

Weight: In drilling fluid terminology, the density of a drilling fluid; typically expressed in either lb/gal, lb/cu ft, psi hydrostatic pressure per 1000 ft of depth, or specific gravity related to water.

Weighted drilling fluid: A drilling fluid to which commercial solids have been added to increase the slurry weight.

Well bore (wellbore): The hole drilled by the bit, also known as the borehole.

Well-bore stabilization: Maintenance of well-bore integrity, which generally requires manipulating the properties of the drilling fluid to simulate the physicochemical environment of the rock before it was drilled.

Wetting agent: A substance that, when added to a liquid, increases the spreading of the liquid on a surface or the penetration of the liquid into a material.

Whipstock: A device inserted into a well bore to cause the drill bit to exit the established path of the existing well bore; the tool used for the initiation of directional drilling.

Wildcat: A well in unproved territory.

Yield point: A term derived from a direct-reading viscometer (Fann V-G or equivalent) based on subtracting the plastic viscosity from the 300-rpm reading; also, an extrapolated shear stress at zero shear rate created by assuming a linear relationship between shear stress and shear rate and determining the intercept on the shear stress axis.

Zero-zero gels: A condition wherein the drilling fluid fails to form measurable gels during a quiescent time interval (usually 10 minutes).

Zeta potential: The electrokinetic potential of a particle as determined by its electrophoretic mobility. This electric potential causes colloidal particles to repel each other and stay in suspension.

Zinc bromide: $ZnBr_2$. A very soluble salt used to increase the density of water or brine to more than double that of water; typically, added to calcium chloride/calcium bromide mixed brines.

Bibliography and Information Sources

Abdel-Aal, H.K., Aggour, A.A., and Fahim, M.A. 2016. Petroleum and Gas Field Processing 2nd Edition. CRC Press, Taylor & Francis Group, Boca Raton, Florida.

Adams, N. 1985. Drilling Engineering: A Complete Well Planning Handbook. Pennwell Publishing Corp., Tulsa, Oklahoma.

Ahmed, T. 2010. Reservoir Engineering Handbook 4th Edition. Gulf Professional Publishing, Elsevier, Burlington, Massachusetts.

ASME. 2004. Drilling Fluids Processing Handbook. Gulf Professional Publishing, Elsevier Science, Burlington, Massachusetts.

Austin, E.H. 1983. Drilling Engineering Handbook. Springer International Publishing AG, Springer International Publishing Cham, Switzerland.

Avallone, E.A., Baumeister, T. III, and Sadegh, A. 2007. Marks' Standard Handbook for Mechanical Engineers 11th Edition. McGraw-Hill Education, New York.

Azar, J.J., and Rebello Samuel, G. 2007. Drilling Engineering. Pennwell Publishing Corp., Tulsa, Oklahoma.

Bahadori, A. 2014. Corrosion and Materials Selection: A Guide for the Chemical and Petroleum Industries. John Wiley & Sons Inc., Hoboken, New Jersey.

Boggs, S., Jr. 2011. Principles of Sedimentology and Stratigraphy 5th Edition. New Jersey, Prentice Hall, Pearson Education Inc., Upper Saddle River, New Jersey.

Branan, C.R. (Editor). 2005. Rules of Thumb for Chemical Engineers: A Manual for Quick, Accurate Solutions to Everyday Process Engineering Problems 3rd Edition. Gulf Professional Publishing, Elsevier Science, Burlington, Massachusetts.

Chaudhuri, U.R. 2010. Fundamentals of Petroleum & Petrochemical Engineering. CRC Press, Taylor & Francis Group, Boca Raton, Florida.

Craft, B.C., Hawkins, M.F. and Terry, R.E. 1991. Applied Petroleum Reservoir Engineering 2nd Edition. Prentice Hall, Englewood Cliffs, New Jersey.

Dake, L.P. 1978. Fundamentals of Reservoir Engineering. Elsevier, Amsterdam, Netherlands.

Darton, R. (Editor). 1997. Distillation and Absorption. Institution of Chemical Engineers, Rugby, United Kingdom. Volume 2.

Devold, H. 2013. Oil and Gas Production Handbook: An Introduction to Oil and Gas Production, Transport, Refining, and Petrochemical Industry 3rd Edition. ABB Oil and Gas, Oslo, Norway.

Evans, F.L. 1979. Equipment Design Handbook for Refineries and Chemical Plants Volume 1. 2nd Edition. Gulf Publishing Company, Houston, Texas.

Fahim, M.A., Alsahhaf, T.A., and Elkilani, A. 2010. Fundamentals of Petroleum Refining. Elsevier, Amsterdam, Netherlands.

Gary, J.G., Handwerk, G.E., and Kaiser, M.J. 2007. Petroleum Refining: Technology and Economics 5th Edition. CRC Press, Taylor & Francis Group, Boca Raton, Florida.

Himmelblau, D.M., and Riggs, J.B. 2012. Basic Principles and Calculations in Chemical Engineering 8th Edition. Prentice Hall, Pearson Education Inc., Upper Saddle River, New Jersey.

Holloway, M.D., and Rudd, O. 2013. Fracking: The Operations and Environmental Consequences of Hydraulic Fracturing. Scrivener Publishing, Beverly, Massachusetts.

Hsu, C.S., and Robinson, P.R. (Editors) 2006. Practical Advances in Petroleum Processing Volume 1 and Volume 2. Springer Science, New York.

Jones, D. 1996. Principles and Prevention of Corrosion 2nd Edition. Prentice Hall, Pearson Education Inc., Upper Saddle River, New jersey.

Lake, L.W. (Editor) 2006. Petroleum Engineering Handbook Volumes I-VII. Society of Petroleum Engineers, Richardson, Texas.

Lavrov, A., and Torsæter, M. 2016. Physics and Mechanics of Primary Well Cementing. Springer international Publishing AG, Cham Switzerland.

Lyons, W.C. (Editor). 1996. Standard Handbook of Petroleum & Natural Gas Engineering, Volumes 1 and 2. Gulf Publishing Company, Houston, Texas.

McCain, W.D. Jr., Spivey, J.P. and Lenn, C.P. 2010. Petroleum Reservoir Fluid Property Correlations. Pennwell Publishing Corp., Tulsa, Oklahoma.

McCain, W.D. Jr. 2017. Properties of Petroleum Fluids 3rd Edition. Pennwell Publishing Corp., Tulsa, Oklahoma.

Mireault, R., and Dean L. (Principal Authors). 2008. Reservoir Engineering for Geologists. Canadian Society of Petroleum Geologists, Calgary, Alberta, Canada.

Mitchell, R.F., and Miska, S.Z. 2011. Fundamentals of Drilling Engineering. Society of Petroleum Engineers, Richardson, Texas.

NIST. 2011. CODATA Recommended Values of the Fundamental Physical Constants: 2006. NIST Standard Reference Database 121. 1 Oct. 2008. Physical Measurement Laboratory of NIST. March 2.

Perry, R.H., and Green, D.W. 1997. Perry's Chemical Engineers' Handbook 7th Edition. McGraw-Hill, New York.

Rachinsky, M.Z., and Kerimov, V.Y. 2015. Fluid Dynamics of Oil and Gas Reservoirs. Scrivener Publishing, Beverly, Massachusetts.

Saldivar Marquez, M.B., Boussaada, I., Mounier, H., Niculescu, S.-I. 2015. Analysis and Control of Oilwell Drilling Vibrations. Springer International Publishing AG, Cham, Switzerland.

Samuel, G.R. 2010. Formulas and Calculations for Drilling Operations 1st Edition. Scrivener Publishing, Beverly, Massachusetts.

Speight, J.G. 2003. Perry's Standard Table and Formulas for Chemical Engineers. McGraw-Hill, New York.

Speight, J.G. 2011. An Introduction to Petroleum Technology, Economics, and Politics. Scrivener Publishing, Beverly, Massachusetts.

Speight, J.G. 2014. The Chemistry and Technology of Petroleum 5th Edition. CRC Press, Taylor & Francis Group, Boca Raton, Florida.

Speight, J.G. 2016. Handbook of Hydraulic Fracturing. John Wiley & Sons Inc., Hoboken, New Jersey.

Speight, J.G. (Editor). 2017. Lange's Handbook of Chemistry 17th Edition. McGraw-Hill Education, New York.

Veatch, R.W., King, G.E., and Holditch, S.A. 2017. Essentials of Hydraulic Fracturing: Vertical And Horizontal Wellbores. Pennwell Publishing Corp., Tulsa, Oklahoma.

Walsh, M.P., and Lake, L.W. 2003. A Generalized Approach to Primary Hydrocarbon Recovery. Elsevier, Amsterdam, Netherlands.

About the Author

Dr. James G. Speight CChem., FRSC, FCIC, FACS earned his B.Sc. and PhD degrees from the University of Manchester, England – he also holds a DSC in The Geological Sciences (VINIGRI, St. Petersburg, Russia) and a PhD in Petroleum Engineering, Dubna International University, Moscow, Russia). Dr. Speight is the author of more than 75 books in petroleum science, petroleum engineering, and environmental sciences. Formerly the CEO of the Western Research Institute and now an independent consultant, he has served as Adjunct Professor in the Department of Chemical and Fuels Engineering at the University of Utah and in the Departments of Chemistry and Chemical and Petroleum Engineering at the University of Wyoming. In addition, he has also been a Visiting Professor in Chemical Engineering at the following universities: University of Missouri-Columbia, Technical University of Denmark, and University of Trinidad and Tobago.

In 1995, Dr. Speight was awarded the Diploma of Honor, (Pi Epsilon Tau) National Petroleum Engineering Society for Outstanding Contributions to the Petroleum Industry. In 1996, he was elected to the Russian Academy of Sciences and awarded the Gold Medal of Honor that same year for outstanding contributions to the field of petroleum sciences. In 2001, the Russian Academy of Sciences also awarded Dr. Speight the Einstein Medal for outstanding contributions and service in the field of Geological Sciences and in 2005 h received the Scientists without Borders Medal of Honor of the Russian Academy of Sciences. In 2006, he was the appointed as the Methanex Distinguished Professor, University of Trinidad and Tobago as well as the Gold Medal – Giants of Science and Engineering, Russian Academy of Sciences, in recognition of Continued Excellence in Science and Engineering.

Index

Also of Interest

Check out these other related titles from Scrivener Publishing

From the Same Author

Rules of Thumb for Petroleum Engineers, by James G. Speight, ISBN 9781118595268. The most comprehensive and thorough reference work available for petroleum engineers of all levels. *NOW AVAILABLE!*

Ethics in the University, by James G. Speight, ISBN 9781118872130. Examining the potential for unethical behavior by all academic staff, both professionals and non-professionals, this groundbreaking new study uses documented examples to show where the matter could have been halted before it became an ethics issue and how to navigate the maze of today's sometimes confusing ethical academic arena. *NOW AVAILABLE!*

Ethics in Engineering, by James G. Speight and Russell Foote, ISBN 9780470626023. Covers the most thought-provoking ethical questions in engineering. *NOW AVAILABLE!*

Peak Energy: Myth or Reality? by James G. Speight and M. R. Islam, ISBN 9781118549421. This groundbreaking study, written by two of the world's foremost authorities in the energy industry, examines our planet's energy future from the perspective of the "peak oil" philosophy. *NOW AVAILABLE!*

Coal-Fired Power Generation Handbook, by James G. Speight, ISBN 9781118208465. The most complete and up-to-date handbook on power generation from coal, this book covers all of today's new, cleaner methods for creating electricity from coal, the environmental challenges and concerns involved in its production, and developing technologies. *NOW AVAILABLE!*

Bioremediation of Petroleum and Petroleum Products, by James Speight and Karuna Arjoon, ISBN 9780470938492. With petroleum-related spills, explosions, and health issues in the headlines almost every day, the issue of remediation of petroleum and petroleum products is taking on increasing importance, for the survival of our environment, our planet, and our future. This book is the first of its kind to explore this difficult issue from an engineering and scientific point of view and offer solutions and reasonable courses of action. *NOW AVAILABLE!*

Made in the USA
Las Vegas, NV
21 September 2026

9d4ab412-3a77-4439-be75-0607188805f4R01